ANNALS *of* THE NEW YORK ACADEMY OF SCIENCES

VOLUME
1305

ISBN-10: 1-57331-906-6; **ISBN-13:** 978-1-57331-906-5

ISSUE

Annals Reports

TABLE OF CONTENTS

T0344848

Annals of the New York Academy of Sciences (ISSN: 0077-8923 [print]; ISSN: 1749-6632 [online]) is published 30 times a year on behalf of the New York Academy of Sciences by Wiley Subscription Services, Inc., a Wiley Company, 111 River Street, Hoboken, NJ 07030-5774.

Mailing: *Annals of the New York Academy of Sciences* is mailed standard rate.

Postmaster: Send all address changes to ANNALS OF THE NEW YORK ACADEMY OF SCIENCES, Journal Customer Services, John Wiley & Sons Inc., 350 Main Street, Malden, MA 02148-5020.

Disclaimer: The publisher, the New York Academy of Sciences, and the editors cannot be held responsible for errors or any consequences arising from the use of information contained in this publication; the views and opinions expressed do not necessarily reflect those of the publisher, the New York Academy of Sciences, and editors, neither does the publication of advertisements constitute any endorsement by the publisher, the New York Academy of Sciences and editors of the products advertised.

Publisher: *Annals of the New York Academy of Sciences* is published by Wiley Periodicals, Inc., Commerce Place, 350 Main Street, Malden, MA 02148; Telephone: 781 388 8200; Fax: 781 388 8210.

Journal Customer Services: For ordering information, claims, and any inquiry concerning your subscription, please go to www.wileycustomerhelp.com/ask or contact your nearest office. *Americas:* Email: cs-journals@wiley.com; Tel:+1 781 388 8598 or 1 800 835 6770 (Toll free in the USA & Canada). *Europe, Middle East, Asia:* Email: cs-journals@wiley. com; Tel: +44 (0) 1865 778315. *Asia Pacific:* Email: cs-journals@wiley.com; Tel: +65 6511 8000. *Japan:* For Japanese speaking support, Email: cs-japan@wiley.com; Tel: +65 6511 8010 or Tel (toll-free): 005 316 50 480. Visit our Online Customer Get-Help available in 6 languages at www.wileycustomerhelp.com.

Information for Subscribers: *Annals of the New York Academy of Sciences* is published in 30 volumes per year. Subscription prices for 2013 are: Print & Online: US$6,053 (US), US$6,589 (Rest of World), €4,269 (Europe), £3,364 (UK). Prices are exclusive of tax. Australian GST, Canadian GST, and European VAT will be applied at the appropriate rates. For more information on current tax rates, please go to www.wileyonlinelibrary.com/tax-vat. The price includes online access to the current and all online back files to January 1, 2009, where available. For other pricing options, including access information and terms and conditions, please visit www.wileyonlinelibrary.com/access.

Delivery Terms and Legal Title: Where the subscription price includes print volumes and delivery is to the recipient's address, delivery terms are Delivered at Place (DAP); the recipient is responsible for paying any import duty or taxes. Title to all volumes transfers FOB our shipping point, freight prepaid. We will endeavour to fulfill claims for missing or damaged copies within six months of publication, within our reasonable discretion and subject to availability.

Back issues: Recent single volumes are available to institutions at the current single volume price from cs-journals@wiley.com. Earlier volumes may be obtained from Periodicals Service Company, 11 Main Street, Germantown, NY 12526, USA. Tel: +1 518 537 4700, Fax: +1 518 537 5899, Email: psc@periodicals.com. For submission instructions, subscription, and all other information visit: www.wileyonlinelibrary.com/journal/nyas.

Production Editors: Kelly McSweeney and Allie Struzik (email: nyas@wiley.com).

Commercial Reprints: Dan Nicholas (email: dnicholas@wiley.com).

Membership information: Members may order copies of *Annals* volumes directly from the Academy by visiting www.nyas.org/annals, emailing customerservice@nyas.org, faxing +1 212 298 3650, or calling 1 800 843 6927 (toll free in the USA), or +1 212 298 8640. For more information on becoming a member of the New York Academy of Sciences, please visit www.nyas.org/membership. Claims and inquiries on member orders should be directed to the Academy at email: membership@nyas.org or Tel: 1 800 843 6927 (toll free in the USA) or +1 212 298 8640.

Printed in the USA by The Sheridan Group.

View *Annals* online at www.wileyonlinelibrary.com/journal/nyas.

Abstracting and Indexing Services: *Annals of the New York Academy of Sciences* is indexed by MEDLINE, Science Citation Index, and SCOPUS. For a complete list of A&I services, please visit the journal homepage at www.wileyonlinelibrary.com/journal/nyas.

Access to *Annals* is available free online within institutions in the developing world through the AGORA initiative with the FAO, the HINARI initiative with the WHO, and the OARE initiative with UNEP. For information, visit www.aginternetwork.org, www.healthinternetwork.org, www.oarescience.org.

Annals of the New York Academy of Sciences accepts articles for Open Access publication. Please visit http://olabout.wiley.com/WileyCDA/Section/id-406241.html for further information about OnlineOpen.

Wiley's Corporate Citizenship initiative seeks to address the environmental, social, economic, and ethical challenges faced in our business and which are important to our diverse stakeholder groups. Since launching the initiative, we have focused on sharing our content with those in need, enhancing community philanthropy, reducing our carbon impact, creating global guidelines and best practices for paper use, establishing a vendor code of ethics, and engaging our colleagues and other stakeholders in our efforts. Follow our progress at www.wiley.com/go/citizenship.

Ann. N.Y. Acad. Sci. ISSN 0077-8923

Evolutionary dynamics and information hierarchies in biological systems

Sara Imari Walker,[1,2,*] Benjamin J. Callahan,[3,*] Gaurav Arya,[4] J. David Barry,[5] Tanmoy Bhattacharya,[6,7] Sergei Grigoryev,[8] Matteo Pellegrini,[9] Karsten Rippe,[10] and Susan M. Rosenberg[11]

[1]BEYOND: Center for Fundamental Concepts in Science, Arizona State University, Tempe, Arizona. [2]Blue Marble Space Institute of Science, Seattle, Washington. [3]Department of Applied Physics, Stanford University, Stanford, California. [4]Department of NanoEngineering, University of California, San Diego, La Jolla, California. [5]Wellcome Trust Centre for Molecular Parasitology, Institute of Infection, Immunity and Inflammation, University of Glasgow, Glasgow, United Kingdom. [6]Sante Fe Institute, Sante Fe, New Mexico. [7]Grp T-2, MSB285, Los Alamos National Laboratory, Los Alamos, New Mexico. [8]Penn State University College of Medicine Department Biochemistry and Molecular Biology, Pennsylvania State University, Hershey, Pennsylvania. [9]Department of Molecular, Cell, and Developmental Biology, University of California Los Angeles, Los Angeles, California. [10]Deutsches Krebsforschungszentrum (DKFZ) and BioQuant, Research Group Genome Organization & Function, Heidelberg, Germany. [11]Departments of Molecular and Human Genetics, Biochemistry and Molecular Biology, Molecular Virology and Microbiology, and Dan L. Duncan Cancer Center, Baylor College of Medicine, Houston, Texas

Address for correspondence: Sara Imari Walker, Ph.D. BEYOND: Center for Fundamental Concepts in Science, P.O. Box 871504, Tempe, AZ 85287. sara.i.walker@asu.edu

The study of evolution has entered a revolutionary new era, where quantitative and predictive methods are transforming the traditionally qualitative and retrospective approaches of the past. Genomic sequencing and modern computational techniques are permitting quantitative comparisons between variation in the natural world and predictions rooted in neo-Darwinian theory, revealing the shortcomings of current evolutionary theory, particularly with regard to large-scale phenomena like macroevolution. Current research spanning and uniting diverse fields and exploring the physical and chemical nature of organisms across temporal, spatial, and organizational scales is replacing the model of evolution as a passive filter selecting for random changes at the nucleotide level with a paradigm in which evolution is a dynamic process both constrained and driven by the informational architecture of organisms across scales, from DNA and chromatin regulation to interactions within and between species and the environment.

Keywords: evolution; information hierarchy; chromatin; epigenetics; viruses; networks

Introduction

The field of evolution is experiencing an exciting period, as it continues to transform from a qualitative and retrospective science into a quantitative and predictive one. Darwin's natural selection and Mendel's genetic inheritance laid the foundation for the development of population genetics and the neo-Darwinian synthesis that followed. But it is now, with the advent of modern technologies—

particularly in the area of sequencing—that we are able to robustly and quantitatively compare the predictions from such theories with the variety of nature. Those quantitative comparisons make clear that large gaps exist between our current understanding of evolutionary processes and what we observe in the natural world. Much of our theory rests on highly simplified caricatures at almost every level of biological organization, from genetic to phenotypic to environmental, and perhaps unsurprisingly such theories run into greater and greater difficulty as we increase our scope. For example, we now understand quantitatively, and have confirmed

---

*These authors contributed equally to this work.

doi: 10.1111/nyas.12140

1

empirically, the evolutionary dynamics of short-term, simple adaptation in (not too large) populations. However, the grander challenges associated with what has been dubbed *macroevolution* largely remain beyond our current quantitative theories, and even short-term *microevolution* confounds us when the real world differs too much from the limited assumptions of our models (e.g., homogeneous populations with random mutations).

During August 2012, a group of physicists and biologists, with diverse backgrounds and representing a broad range of research interests, came together for a workshop on "Evolutionary Dynamics and Information Hierarchies in Biological Systems" at the Aspen Center for Physics in Aspen, Colorado to discuss these challenges and to explore new approaches. The three themed weeks of the workshop focused on the organization of DNA into chromatin, epigenetic adaptation and host/pathogen interaction, and macroevolution. Although these areas represent a wide breadth of biological phenomena, several unifying themes emerged through workshop discussions. In particular, the differences between the simplicity of our theoretical models and the complex interactions characteristic of real *physical* systems were repeatedly highlighted. Workshop discussions therefore pointed to key areas where theory and observations should aim to converge as we refine our understanding of evolution.

Among the starkest contrasts between theory and reality emerged through dialogue on the difference between the physical DNA molecule and the disembodied string of letters by which we represent it. The physical structure of DNA, how it is packed and twisted and altered, is not captured by a simple string of letters, but is vital to its function. DNA in eukaryotes typically exists in the form of chromatin, a condensed network of DNA and protein, the structure of which influences the interpretation of the string of letters. Chromatin sits in a mesoscopic regime that controls the flow of information between the microscopic nucleotides and the macroscopic phenotypic traits of the cell. And the physical and chemical heterogeneity of DNA begets heterogeneity in the mutation process, making it possible for evolution to act on the mutational process even as it is fueled by it.

A second recurring contrast was apparent when comparing the well-mixed populations, usually of fixed size, that dominate our models, and the heterogeneous population regimes of nature. This difference was vividly illustrated in discussions about host–pathogen dynamics where large intra-host populations compete against aggressive immune systems, but where the broader dynamics across hosts depends on relatively low numbers and low rates of dispersal between hosts. The interplay between these two levels of population structure leads to richer and more complex pathogen dynamics than is permitted by traditional models. Further discussions centered on the influence of population structure: even when a single population regime is appropriate, the physical and network structure of real populations can dramatically change evolutionary outcomes.

The final emerging contrast at the workshop, and perhaps the most profound, was the fundamental disconnect between our flat evolutionary models that typically focus on one biological length and/or time scale, and the multi-scale hierarchies characteristic of living organisms. For example, life organizes information in a complex hierarchy ranging from DNA sequences and chromatin regulation to cellular signaling and tissue/organ organization, and to the interactions between organisms and species. All levels of this hierarchy influence fitness, and therefore selection acts simultaneously on a variety of scales ranging from the microscale (e.g., DNA) to the macroscale (e.g., ecosystems). Evolution does not act at each scale in isolation. Workshop discussions therefore focused on connecting diverse aspects of this informational hierarchy in biological systems, and how the connections between multiple temporal and spatial scales interplay with the evolutionary process. Key questions that repeatedly arose from these discussions were: What is the structure of the biological networks that transforms chemistry (e.g., genomes, protein networks) into living organisms? And how might that structure constrain or facilitate evolution? As discussed at the workshop, answers to these questions may have implications for our understanding of the emergence of life.

These workshop themes point to the importance of novel cross-disciplinary approaches in bridging the gap between our simplified models of evolutionary processes and the reality of living organisms as physical and chemical entities. The chromatin structure of DNA requires us to go beyond sequencing to characterize it, and beyond the sequence to represent it. The evolution of pathogens requires us to account for both their population dynamics within a host and their epidemiology across a population of hosts. A deep understanding of life cannot end with a catalogue of variation, but must also describe the framework in which variation emerges and is processed. Workshop discussions repeatedly highlighted how accounting for the physical and chemical nature of organisms at various temporal, spatial, and organizational scales can lead to new perspectives on evolution. The resulting paradigm shifts may substantially differ from the picture of evolution as a passive selective filter acting on random variation provided by the neo-Darwinian synthesis. While the workshop overview provided here cannot be an exhaustive review of such a broad range of topics, it highlights these emerging themes and the open challenges that arose through workshop discussions, and references the individual contributions of workshop participants for more in-depth discussion.

Randomness and evolvability

Several workshop discussions focused on the need for an expanded picture of evolutionary processes that goes beyond the neo-Darwinian synthesis. Neo-Darwinism, synthesizing Darwinian evolution through natural selection with Mendelian genetics, asserts that mutation is a purely random event in genomic space and blind to selective environments. However, this picture is inconsistent with biochemical reality, suggesting that a deeper understanding that integrates physical and chemical insights is required. Workshop presentations on this topic explored experiments demonstrating that some mutations are not random with respect to DNA sequence, to time, or to their potential effect on survival. In particular, emphasis was placed on the role of stress-induced mutagenesis, mutational hotspots, and the feedback between an organism and its environment in natural selection. During the discussions many fascinating questions emerged about variation in adaptive potential as a function of both environment and stress, pointing to a more dynamic picture of the mutational process than that assumed by the neo-Darwinian synthesis.

Mutation lacks foresight, but it can have hindsight

The theory of evolution states that members of natural populations vary in many ways, and that selection favors inheritance of those traits most fitted to the environment. As evolution is currently taught, new variants of genes are generated by mutations that are random with respect to their probability of being adaptive. Lynn Caporale (St. John's University) presented evidence that the assumption that "all mutation is random" is not consistent with a growing body of data;[1] she also asserted that the statement that all mutation is random is not actually consistent with the theory of evolution.

The assumption that mutations must be random with respect to adaptive value arises from the argument that processes that generate mutation have "no foresight." This would be true if environmental change was random, yet many challenges recur, such as the need to combat pathogens (and the need for pathogens to avoid host defenses). A lineage that evolves an effective response to a class of challenges would be expected to survive such repeated challenge more effectively than one that responded randomly each time.

Due to sequence-dependent variations in physical chemical properties, the probabilities of distinct classes of mutation vary along the DNA sequence. Since selection acts on variation, Caporale pointed out that evolutionary theory actually predicts that selection can act to make mutations non-random with respect to their potential effect on survival. An environment that changes in non-random ways selects for non-random variation.[2] Caporale then described multiple examples of non-random mutation. For example, pathogens with non-codon length repetitive sequences (such as CAATCAAT-CAATCAAT) in their coat genes generate immune-defying coat protein variants at rates 1000 times that of the background mutation rate.[3]

One widely used protocol that enables efficient storage of extensive diversity is the use of fragments of genes with identifying tags recognized by other genes or gene products. This protocol effectively encodes rules for the assembly of a diverse set of genes not explicitly encoded in the genome.[2,4] Among systems discussed at the Aspen workshop in which the use of gene fragments and rules for their assembly has evolved to encode diversity are the vertebrate immune system and trypanosome coat proteins (see e.g., the section highlighting Dave Barry's work below).

DNA sequence variation can be regulated biochemically in many ways: through altering nucleotide pools, decreasing mismatch repair, changing the balance among different repair proteins, inducing novel polymerases, and releasing transposable elements. Caporale explained that in contrast to widely used statements of evolutionary theory, we should not assume that mutation is constant in time. One clear example of such temporal change is the increased mutation rate that can accompany stress (defined as a sensed maladaptation to the environment that results in the activation of a biochemical stress response) that was presented at the workshop by Susan Rosenberg.

Mutation as a stress response and the regulation of evolvability

Susan Rosenberg (Baylor College of Medicine) presented experimental evidence that mutagenesis is regulated in both *time*, through increased rates of mutagenesis during periods of stress, which generates new mutations specifically when cells or organisms are maladapted to their environments, and in genomic *space*, in which mutations are observed to cluster in genomes.

Although in unstressed *Escherichia coli* cells repair of double-strand breaks by homologous recombination is non-mutagenic and uses high-fidelity DNA polymerase III (Pol III), when stressed cells switch to a mutagenic mode of DNA break repair that uses error-prone DNA polymerase DinB. DinB participates in DNA break repair and generates mutations under control of the SOS DNA damage response and the RpoS-general/starvation stress response.[5,6] Rosenberg's data suggest that most spontaneous mutation in starved *E. coli* results from DNA double-strand break (DSB)–dependent stress-induced mutagenesis (SIM), requiring (1) DSBs and their repair by recombination, (2) activation of the SOS DNA damage response, which upregulates DinB levels, and (3) a separate stress such as starvation that activates the RpoS general stress response, which allows DinB DNA polymerase to participate in mutagenic break repair.[6] Interestingly, Rosenberg showed that artificial activation of the stress-response is sufficient to trigger SIM even in unstressed cells (i.e., stress itself is not required).

Various studies had previously suggested that genomic mutation hotspots might be related to DSBs,[5,7–10] but the results were open to multiple interpretations. By engineering DSBs at various sites in the genome, Rosenberg's team found that DSBs pro-

duce two distinct kinds of mutation hotspots that form by different mechanisms.[11] The first are strong local hotspots that are maximal within the first few kilobases (and extend to 60kb) of repaired DSBs, and which form via RecBCD-dependent exonucleolytic resection from DSBs and gap-filling synthesis. The second are weak long-distance hot zones extending up to approximately 1 Mb away from the DSB site, and that form independently of resection, probably via break-induced replication. That mutations are confined to local zones by the coupling of mutagenesis to DSB repair could be evolutionarily important in that it allows multiple simultaneous mutations within genes.

Rosenberg concluded with the identification of a large protein network required by *E. coli* to run the program of mutagenic repair of DNA breaks in response to stress.[12] The 93 genes identified as either promoting or required for stress-induced mutagenesis include 21 regulatory genes, 7 proteases/chaperones, 12 genes involved in DNA replication and repair, 20 genes that encode electron transfer functions, 8 genes involved in metabolism, 12 in cellular processing, and 12 of unknown function. More than half of the genes sense stress and transduce the stress signal that ultimately allows activation of three critical stress response regulators, which are key network hubs: the SOS DNA damage response, the RpoS general stress response, and the RpoE membrane protein stress response. The surprising conclusions of this study are that (1) many proteins are required to run a relatively simple program of mutagenic DNA repair; (2) most of the proteins function in stress signal transduction to key network hubs—the stress-response activators, which allow mutagenesis; and therefore (3) the large number of proteins allocated to sensing and communicating stress makes it clear that increasing mutagenesis at times of stress is important. Understanding how individual genes within this network affect stress-induced mutation is shedding important light on how SIM affects system-level evolution of protein networks, thus providing important insights into the hierarchy of control of evolvability at a systems level.

Epigenetics and chromatin organization

The discussions on the evolvability of mutation rates (discussed above) highlighted the importance of the physical and chemical structure of DNA in understanding evolutionary processes. In eukaryotes, nucleic DNA is organized into chromatin—a highly condensed complex of DNA and protein. The structure of chromatin helps determine which stretches of DNA are read, and when, and can also influence fundamental evolutionary processes such as mutation rates.[13] However, much work remains to understand the physical structure of chromatin, especially *in vivo*, and how it is epigenetically regulated. The multiple spatial and temporal scales involved in the hierarchical processes by which the eukaryotic genome is packaged into chromatin complicate these efforts. Several workshop presentations explored current research on chromatin structure and the way in which epigenetic networks control access to chromatin. The results presented illuminate the constraints that epigenetic regulation and the physical structure of chromatin impose on evolution, thus extending our understanding of how information processing within the eukaryotic cell influences adaptive processes.

Modeling DNA organization in nucleosomes, chromatin, and chromosomes

Gaurav Arya (University of California, San Diego) presented his group's recent efforts in developing computational models that can describe the packing of DNA into chromatin at the multiple length scales at which this structuring happens. Arya began by reviewing the fascinating hierarchical process by which the eukaryotic genome is packaged. He then stressed the need to understand the organization within each hierarchical level due to the critical role of chromatin organization in modulating DNA accessibility, protein binding, and long-range genomic interactions, as well as its more apparent role in genome packaging. He particularly emphasized the importance of coarse-grained (CG) models for developing such an understanding, i.e., models capable of probing the large length and time scales of each organizational level that cannot be probed by atomistic models.

At the lowest level, eukaryotic chromatin is composed of repeats of structurally uniform units called nucleosomes, which consist of DNA segments of approximately 147 base pairs wrapped around an octamer of histone proteins. Arya developed a CG model that elucidates the dynamics of force-induced unraveling of nucleosomes, a problem relevant to DNA accessibility and nucleosome remodeling.[14]

In this model, the DNA and histone octamer are treated as separate entities capable of assembling and disassembling. DNA–histone interactions are parameterized to reproduce the DNA–histone interaction free energy profile and unwrapping forces obtained from single-molecule experiments. Arya used Brownian dynamics simulations of the nucleosome where the flanking DNA were pulled apart at fixed speeds to explore the dynamics of the wrapped DNA and the motions of the histone octamer accompanying nucleosome unraveling. An important finding is the role that non-uniform DNA–histone interactions along wrapped DNA play in stabilizing nucleosomes against unraveling, while enhancing the accessibility of the wound DNA via breathing motions.

Moving beyond individual nucleosomes, Arya, in collaboration with Tamar Schlick, developed a CG model of nucleosome arrays—segments of bound nucleosomes connected by a contiguous strand of DNA—that describes their conformation and interactions using a few important degrees of freedom, while accounting for key physical features including thermal fluctuations, configurational entropy, DNA mechanics, nucleosome shape, linker histone binding, histone tail flexibility, excluded volume, and salt-screened electrostatic interactions.[15,16] Arya's Monte Carlo simulations of this model demonstrate a polymorphic structure of chromatin fibers, which fits well with the crosslinking experiments of Sergei Grigoryev (discussed in the next section), and which reveal the importance of physiological salt, histone tails, and the linker histone to the stability of the compact state of chromatin at this level of packaging.

Finally, motivated by the need to understand higher-order chromatin folding, Arya described a computational approach to determine chromatin conformations from interaction frequencies (IFs) measured by chromosome conformation capture and related techniques.[17] Dynamic simulations of a restrained bead-chain model are used to estimate the IFs. These estimates are then incorporated into an adaptive algorithm that iteratively refines the strengths of the imposed restraints on the bead-chain until a match between the computed and experimental IFs is achieved. This approach has been validated against multiple simulated test systems and is currently being refined against experiments.

Higher order chromatin folding

Sergei Grigoryev (Pennsylvania State University College of Medicine) delved further into higher-order chromatin folding, focusing on the role of the nucleosome repeat length (NRL) and nucleosome–nucleosome interactions. Nucleosome arrays fold into higher-order chromatin, which controls chromatin condensation and its accessibility for transcription, recombination, and other DNA-directed biological processes. However, as highlighted above, the architecture of higher-order chromatin is still poorly understood, and many of its physical properties are unknown.

Over recent years, Grigoryev's laboratory has extensively characterized condensed chromatin in several types of terminally differentiated cells from chicken, mouse, and human. Grigoryev discussed observations that chromatin condensation in each cell type was associated with changes in NRL, the concentrations of linker histone and tissue-specific non-histone architectural proteins, as well as post-translational histone modifications. These studies have revealed significant differences between organisms and tissue types, suggesting that the process of developmental regulation of chromatin condensation, although fundamental for cell differentiation, is not evolutionary conserved. Experiments also suggest that in some cell types, chromatin condensation may involve massive interdigitation between folded nucleosome arrays promoted by chromatin bridging factors and histone modifications that mediate nucleosome interactions between the arrays.

Grigoryev next focused on the contribution of the NRL to chromatin higher-order structure, as revealed by work on synthetic DNA sequences which vary only in the spacing between repeated nucleosome-specific DNA sub-sequences (i.e., their NRL).[18] High-resolution nuclease mapping of these sequences showed that nucleosome arrays maintain protection of DNA from nuclease activity by linker histones, consistent with formation of linker DNA stems observed by electron microscopy (EM). The use of sedimentation and EM techniques revealed an overall negative correlation between NRL and chromatin folding. In the shorter NRL range of 165–177 base pairs (typical of less condensed, transcriptionally-active chromatin), Grigoryev described a strong periodic dependence of chromatin folding on small changes in NRL. This relation-ship suggests that the transcriptionally active yeast genome might have evolved to have precise, short NRLs (162 and 172 base pairs) supporting relatively open higher-order structure. In contrast, the longer NRLs (188 base pairs and above) typical of vertebrate chromatin do not affect chromatin folding and need additional architectural to mediate chromatin condensation.

Grigoryev also presented studies of internal nucleosome interaction within reconstituted and native chromatin using an EM-assisted nucleosome interaction capture (EMANIC) technique.[15] For native and reconstituted chromatin condensed at physiological conditions, the experiments revealed a nucleosome interaction pattern consistent with predominantly straight linkers and a two-start helical arrangement of nucleosome cores. For the most condensed chromatin in the nuclei of terminally-differentiated cells and metaphase chromosomes, Grigoryev also discussed observations of a prevalence of nucleosome interactions typical of the two-start helix, but observed interactions also included those from folded and interdigitated chromatin fibers. The findings were discussed in relation to the mechanism of nucleosome array folding mediated by dynamic short-range nucleosome interactions, which occur even in the most condensed chromatin state.

An epigenetic mechanism to silence transcription in heterochromatin

Karsten Rippe (German Cancer Research Center) described how linking epigenetic modifications of histone residues with their readout by specific protein domains is an important aspect of current theoretical models that describe epigenetic networks.[19,20] These models are frequently characterized by a combination of feedback loops to establish bistable chromatin states: for the locus under consideration, two distinct chromatin states can stably coexist for a certain set of conditions. With respect to quantitative descriptions of epigenetic networks, three fundamental questions are particularly relevant: (1) How is the separation of the genome in active and silenced chromatin states established and maintained and what are the factors that provide specificity for distinct chromatin states? (2) How is the confinement of a given chromatin state to a certain genomic locus achieved? (3) How is a given chromatin state transmitted through the

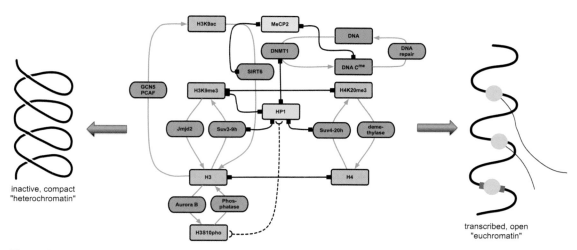

Figure 1. Regulatory epigenetic network that operates at mouse pericentric heterochromatin to silence transcription from satellite repeats. Dependent on the degree of histone H3 trimethylation at lysine 9 (H3K9me3), the system can undergo transitions between a biologically inactive heterochromatin state (high H3K9me3 levels) and an open euchromatin conformation competent for transcription (low H3K9me3 levels). The H3K9me3 modification is set by the Suv3–9h methyltransferase and is recognized by heterochromatin protein 1 (HP1). Demethylation of H3K9me3 by Jmjd2 and histone acetylation promote activation and chromatin opening. Note the linkages between the different epigenetic modifications histone trimethylation, histone acetylation and DNA methylation via protein–protein interactions of network components. Color code: red, chromatin-modifying enzymatic activities; blue, histones, with their posttranslationally methylated (me), acetylated (ac) and phosphorylated (pho) states at the indicated residues; green, DNA; yellow, structural chromatin components. The solid black lines symbolize association between proteins, while the dashed line indicates inhibition of interaction.

cell cycle, i.e., how does the cell's epigenetic memory work?

To address these questions, Rippe introduced his group's work on a particular epigenetic mechanism that silences transcription of repetitive sequences found at the pericentromeric regions (near the center of a chromosome) of the genome in mouse fibroblast cells[21] (Fig. 1). The pericentric heterochromatin (PCH) state is characterized by DNA cytosine methylation and the trimethylation of histone H3 at lysine 9 (H3K9me3) and histone H4 at lysine 20 (H4K20me3), as well as the enrichment of several chromatin modifiers and additional protein components. Studying this system to dissect epigenetic networks has the particular advantage that the PCH domains can be readily identified on fluorescence microscopy images as chromatin-dense foci. Accordingly, Rippe and colleagues were able to compare the dynamic features of PCH with the bona fide biologically active euchromatin (a state of lightly packed chromatin) that surrounds PCH by applying a previously established framework for an integrated fluorescence microscopy–based bleaching and correlation analysis in single living cells.[22,23]

With this technique, they quantified protein concentrations as well as protein–chromatin and protein–protein interactions of the core protein components of PCH. The resulting comprehensive data set was used to model the epigenetic network that is active in PCH.

Rippe also discussed his group's mechanistic analysis of PCH in mouse fibroblasts as a prototypic system to explain how a repressive heterochromatin state is established and maintained. A particularly interesting result from their work is the development of a model in which the H3K9me3 modifications in PCH originate from sparsely distributed nucleation sites that distribute this modification via looping of the nucleosome chain to nucleosomes in spatial proximity. Based on previous studies, the collision probability between nucleation sites and surrounding nucleosomes was converted into a concentration[24–26] that, when cast into a model that incorporated the experimentally determined interaction parameters, yielded an excellent agreement with the measured features of mouse PCH. These results provide steps toward a quantitative understanding of epigenetic activation and silencing and how epigenetic states are maintained.

Using high throughput sequencing to measure genome-wide chromatin structure

Matteo Pellegrini (University of California, Los Angeles) described the insights his group has gained into chromatin structure across entire genomes through next-generation sequencing technologies. A genome-wide analysis of nucleosome positions in the small flowering *Arabidopsis* plant[27] revealed that nucleosomes are enriched in certain areas of the genome, in particular in exons, and especially exon–intron boundaries. This pattern correlates with the enrichment of RNA polymerase II (Pol II) and DNA methylation in exons, consistent with a nucleosomal role in regulating Pol II processivity[28] and the targeting of DNA methylation to nucleosomes, respectively. Augmenting the nucleosome positional data with genome-wide profiles of DNA methylation demonstrated that nucleosome-bound DNA is more methylated than flanking DNA, and revealed 10-bp periodicities in the DNA methylation status of nucleosome-bound DNA. These results indicate that nucleosome positioning influences DNA methylation patterning throughout the entire genome and that DNA methyltransferases preferentially target nucleosome-bound DNA.

DNA methylation and nucleosome densities play a critical role in the regulation of gene expression,[29] but little is known about the degree to which they contribute to the differences among tissues. Pellegrini next presented tissue-specific data in which DNA methylation, nucleosome densities, and transcriptional levels were compared across tissue types. Results showed that nucleosome density is correlated with methylation and inversely correlated with gene expression. A group of root-specific genes was identified that appears to be an example of differential regulation by epigenetic marks—they are preferentially methylated, have lower nucleosome density, and at least tenfold higher expression in *Arabidopsis* roots relative to shoots—supporting a role for chromatin structure in tissue determination.

Finally, Pellegrini discussed work investigating chromatin structure on the megabase scale to identify long-range chromatin conformations, using genome-wide chromosome conformation capture (3C) coupled to high-throughput sequencing (4C-seq) to define the DNA–DNA contacts (*interactomes*) made by a number of genetic loci in pluripotent and differentiated cell types.[30] This technique enabled the identification of long-range DNA–DNA interactions that were confirmed with fluorescence *in situ* hybridization (FISH) and reciprocal 4C-seq. The data showed an organization of long-range DNA–DNA interactions specific to embryonic stem cells that is lost upon differentiation and re-established during reprogramming of differentiated cells to the induced pluripotent state. The genomic features at a given locus (i.e., transcription factor binding and chromatin state) correlate strongly with the genomic features in that locus's interactome (its entire set of molecular interactions). Pellegrini concluded by presenting evidence that chromatin could act combinatorially to guide long-range DNA–DNA interactions that account for the differences between the interactomes of pluripotent and differentiated cells, thus providing insights into epigenetic mechanisms at the level of entire genomes.

The interplay between diversity, function, and the evolution of networks

A unifying theme throughout the workshop was connecting evolutionary processes on multiple scales, ranging from the bacterial and eukaryotic genomes highlighted in the previous sections to gene networks, species and ecosystems, and even technological systems. The broad scope of these discussions encompassed a lively public panel dialog featuring workshop participants Sergei Maslov and Kim Sneppen on the topic of "Randomness and Selection in Biological and Technological Evolution", which was held the evening of August 23. The panel covered the interplay between function and randomness in examples of both biological and technological networks and stimulated much discussion among audience members including scientists and the public. Talks during the workshop also expanded on this theme. Topics covered included the roles of popularity and function in determining selection outcomes in biological (bacterial genomes) and technological (Linux installations) systems, and the evolution of networked populations in biological systems in the context of the generation and maintenance of diversity. An interesting thread connecting these diverse systems that emerged through workshop discussion is the role of network diversity and structure (both population and spatial) in shaping evolutionary processes.

Why networks make gene families like Linux packages

Sergei Maslov (Brookhaven National Lab) explored a network-inspired analogy between the structure of biological and technological systems by comparing bacterial genomes and Linux installations. In bacteria (and Linux alike), each genome (installation) contains only a subset of the much larger universe of orthologous gene families (software packages) that are available to them by horizontal gene transfer (download). The specific subsets of the potential components observed in genomes (installations) are then the product of selection at the whole-organism (whole-computer) level for overall function, by nature (or by the user).

Maslov quantified this analogy by comparing *component frequency distributions*, defined for bacterial genomes (Linux installations) as the number of orthologs (packages) that are present at a given frequency across genomes (installations). Results from ~500 fully sequenced bacterial genomes[31] were compared with those from ~2 million Linux installations (Ubuntu Popularity Contest). In both cases the structure of the component frequency distribution can be broken into three distinct segments: (1) a large "cloud" of low frequency (present in <5% of genomes or installations) components, with numbers increasing quickly as frequency approaches zero, (2) a "shell" of intermediate frequency components, with numbers slowly decreasing as frequency increases, and (3) a small "core" of high-frequency (>90%) components found in most genomes or installations. Excluding the "core", this distribution was well fit by a power-law with an exponent of approximately −1.5, i.e., the probability (P) of an ortholog (package) being found with frequency (f) across all genomes (installations) is $p(f) \sim f^{-1.5}$.

The "cloud" and much of the "shell" consist of genes or packages that implement a variety of features, but whose function requires the presence of other genes or packages. This set of relationships can be represented in network form.[32] At the base of this dependency network are the basic and universal functions of the "core", e.g., RNA polymerase in the case of bacterial genomes or *gcc* in the case of Linux packages, with functions of increasing specificity and complexity being implemented on top of the genes or packages on which they depend (and are therefore connected to in the dependency network). Using data on the dependencies of Linux packages, Maslov showed quantitative agreement between the observed component frequency distribution and that predicted by the known dependencies among Linux packages. Maslov demonstrated that the component frequency distribution we observe across genomes (installations) is a function of the statistical properties of these dependency networks, in particular the average number of dependencies per gene (package). The results suggest that the concordance between the component frequency distributions in bacterial genomes and Linux distributions might represent a concordance in the underlying dependency networks of both bacterial genes and Linux packages, suggesting some universal organizational principles may be at work.

The origins and maintenance of diversity

Kim Sneppen (Niels Bohr Institute) discussed the role of both network and spatial structure in supporting species diversity, addressing the question: What are the minimal requirements for maintaining species diversity over long time scales? Sneppen made use of lichen ecology as a model system suitable for exploring this broad question. Lichens are organisms consisting of a symbiotic union of fungi and algae[33] that primarily inhabit surfaces. When two crustose lichen species interact on a surface, a contact boundary is formed. Diversity is maintained when these boundaries are formed between competitively equal species. Therefore, bulging boundaries observed in lichen communities suggest complex dynamics where one species may overtake another.

The model system described by Sneppen consists of a community of species competing on a two-dimensional lattice, with the ecosystem characterized by a directed network of possible species interactions.[34,35] Since species are spatially distributed, not all interactions between extant species are physically realized at a given time. The average number of species present (D) is determined by the average fraction of species that are invasible by another species (when neighbors), parameterized by γ. Sneppen introduced new species to the community at a rate α, and observed the outcome at the level of overall species diversity. The system displays a first-order phase transition at $\gamma = \gamma_c = 0.055$, transitioning from a low-diversity ($D \sim 1$) to a high-diversity ($D \sim 20$) state as the interaction probability γ is decreased from 1 in the limit of $\alpha \to 0$.[34] Near the critical point at γ_c, the system displays bistability between low and high diversity states, where the transition between states is triggered by fragmentation of populations into patches.[35] Interestingly, these patches act as seed sites for new species to nucleate and spread.

In his discussion, Sneppen highlighted many interesting facets of the observed dynamics in this model system. One important observation was the complete collapse of population structure for systems with random neighbors (such that interactions are not determined by spatial proximity), indicating that spatial organization plays an important role in generating and maintaining diversity. Other limits discussed included introduction of long-range migration and random deaths, which both led to the loss of diversity. The results suggest a complex dynamic between network and spatial structure that dictates species diversity, and leaves many open questions. Among these, it is unclear what determines the length-scale cut-off for patch-size. From an evolutionary perspective, a particularly interesting facet of the models discussed by Sneppen is that they do not rely on a predefined fitness landscape. Instead, the fitness of an individual species is determined dynamically by its interactions with the local networked community within which it is embedded.

Coevolution and the evolutionary arms race

The co-evolutionary arms race between pathogens and the adaptive immune systems of mammals, particularly humans, was the subject of several presentations and much informal discussion at the workshop. The intense pressure applied by the mammalian immune system has driven remarkable architectural changes in the pathogens that must cope with it, as dramatically illustrated by Dave Barry in the case of *Trypanosoma brucei*—the parasitic cause of sleeping sickness. The implications of the combination of the fast and intense in-host arms race with slower dynamics at larger scales, such as the worldwide flu pandemic, and longer times, such as observed in the chronic stage of HIV, were also discussed, with a focus on current challenges in understanding and predicting the evolution of pathogens experiencing these heterogeneous regimes in space, time, and population structure.

The combinatorial diversity arsenal of the sleeping sickness parasite

Dave Barry (University of Glasgow) introduced the anti-immune system of *Trypanosoma brucei,* the single-celled protozoan that causes African sleeping sickness in humans. Trypanosomes have a dense coat of variant surface glycoprotein (VSG), which physically shields them from antibodies against conserved surface proteins and innate immunity mechanisms.[36] The only target for the immune system is VSG itself, but this is a quickly shifting target over the course of infection due to rapid switching of expression among VSG genes.[37] Trypanosomes have a very large potential for variation as VSG seems to have no function other than physically coating the cell.[38] Many VSG variants, with differences of often 80% or more at the amino acid level, appear over the course of an infection. This variation is encoded in the genome as an "archive" of thousands of VSG genes, most of which are non-functional (pseudogenes).[39] These genes sit in the mutable subtelomeric regions near the end of chromosomes, and up to 200 can be found in mini-chromosomes that contain only *VSG* genes.[40] Singular expression of VSG by each trypanosome is achieved by transcription being restricted to only a few loci, to which genes must move to become active.

Switching, which changes which VSG is expressed at a rate of $\sim 10^{-3}$ per generation, involves a process possibly initiated by a DSB at the expression site, followed by DNA repair–mediated replacement of the expressed gene with a copy of any accessible VSG variant in the genome.[41] In trypanosomes this gene conversion process often results in mosaic genes, pieced together from stretches of more than one genomic VSG variant. Thus, even pseudogenes can contribute, and the combinatorial nature of their involvement compounds with the already staggeringly large diversity that is latent within each trypanosome genome.

Archive VSG genes mutate at a high rate, using processes of gene duplication, base substitution, insertions and deletions, and segmental conversion. Their elevated mutation rate appears to be the result of second-order selection: sequence analysis shows that the subtelomeres, in which VSG genes are primarily located, mutate several-fold faster than chromosome cores. Indeed, subtelomeres are increasingly being seen as havens for diversification of eukaryotic multigene families that encode hyper-variable phenotypes. How trypanosome cores and subtelomeres achieve fundamental mutational differences might be linked to epigenetic differences, such as base modification[41] and binding of the ORC1 replication protein.[43,44]

Despite the stochastic disorder underlying VSG switching, antigenic variation is structured. For example, variants are expressed in partially predictable order. Organization is thought to be essential for trypanosomes to deal effectively with the primary sources of selective pressure—the immune system. Modeling studies have recently begun to reveal a network of interactions from the trypanosome genome, through population dynamics, to host populations in the field. This integrative approach engenders not only predictions of how pathogen population processes are determined by underlying molecular genetics, but also inferences about resulting selective pressures on antigen gene archives.

Influenza: genomics of a "Red Queen's race"

Michael Lässig (University of Cologne) discussed the rapid evolution of seasonal influenza A virus (H3N2). Influenza evolves at a rate 10^5 times greater than *Drosophila*, and approximately 25% of nucleotides in the influenza virus have mutated since 1968. Lässig described how such rapid evolution results from immune selection that drives an evolutionary arms race between the pathogen and its host. A striking feature of this process is its punctuated pattern. Adaptive changes occur primarily in antigenic epitopes, i.e., the antibody-binding domains of the viral hemagglutinin. This process involves recurrent selective sweeps, in which clusters of simultaneous nucleotide fixations in the hemagglutinin coding sequence are observed about every four years, with a corresponding drop in diversity. The evolutionary origins of this pattern remain controversial.

Lässig suggested that the rapid adaptation of the influenza A virus produces clonal interference within the hemagglutinin gene that results in the observed punctuated pattern resulting from recurrent selective sweeps of the population.[45] Influenza A might therefore undergo a mode of evolution driven by a "Red Queen's race" between viral strains with different beneficial mutations. To infer selection under clonal interference, Lässig introduced a new method that relies on two measures: a frequency propagator ($G(x)$) defining the likelihood that a new allele reaches a frequency larger than x at some later time, and a loss propagator ($H(x)$) defining the likelihood that a new allele reaches frequencies exceeding a given threshold x at some intermediate point of its lifetime but is eventually lost. By evaluating either nonsynonymous ($G(x)$ and $H(x)$) or synonymous ($G_0(x)$ and $H_0(x)$) mutations for a sample of influenza genome sequences taken over the past 39 years, Lässig described identifying the presence of clonal interference if two conditions are fulfilled: $G(x) / G_0(x) > 1$ and $H(x) / H_0(x) > 1$ for intermediate and large frequency x. The first condition signals predominantly positive selection for a class of mutations, and the second indicates interference interactions: new beneficial alleles rise to substantial frequency, but are eventually driven to loss by a competing clone. Lässig showed that the data for influenza are consistent with clonal interference, with at least one strongly beneficial amino acid substitution per year, where a given selective sweep has on average three to four driving mutations.

In the discussion of the broader implications of this work, Lässig noted that the results strongly suggest that the course of influenza evolution is determined not only by antigenic changes, but also evolutionary competition and selection among viral strain variants: successful viral strains are those that maximize the total fitness of antigen–antibody interactions and of other viral functions by a joint process of adaptation and conservation. He also noted that calculations of the fitness flux[46] suggest that an increase in immune challenge would strongly compromise the viability of influenza. Thus, Lässig concluded that while antigenic adaptation has been a focus of influenza research so far, a broader picture of viral function and fitness is needed to understand the processes shaping influenza evolution.

Viral evolution in response to immunity

Tanmoy Bhattacharya (Sante Fe Institute and Los Alamos National Lab) discussed how viruses evolve to escape immunity, using HIV as a case study. Simian immunodeficiency viruses (SIVs), such as HIV, are lentiviruses (retroviruses with a long incubation period, capable of infecting non-dividing cells) that have coevolved with their hosts for millions of years[47] and rarely produce acute diseases in their natural hosts, but can be pathogenic in other hosts.[48] HIV jumped to humans from chimpanzees in the Congo region of Western Africa[49] around

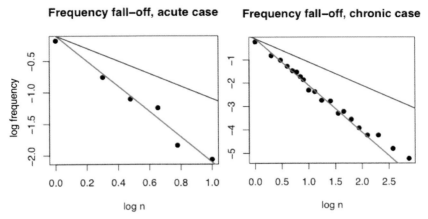

Figure 2. Both the exponentially growing HIV population and acute infection and the constant-sized population from a chronic infection show the characteristic inverse-square fall-off of number of clones with a certain clone size. This indicates repeated rapid population replacements even during the chronic phase.

100 years ago,[50] and is responsible for AIDS and death.[51,52] After an initial period of slow growth, the epidemic has been growing exponentially since the 1960s.[53] During this period, it broke up into roughly geographically separated subtypes, with distinct genetic signatures inherited from the clade founders.

With a mutation rate of about one change per genome every three replications,[54] a high reproductive number of about 10,[55] and a short generation time of about two days,[56] HIV is expected to adapt very quickly to its environment. Early indications that the majority of the sites in the HIV genome were under intense and consistent selective pressure from the host immune system[57] turned out to be based on an incorrect handling of the ancestral genetic differences between the subtypes.[58] In fact, the diversity of the human immune system makes only very few sites display overt adaptation signatures in population-level data.[58] Nevertheless, deep sequencing data provide indications of unexpectedly early and fast escapes from host immune reaction, often resulting in fixation of escape mutations on the time scale of a few weeks.[59]

This difference between a diversifying evolution in the population over 100 years and the "Red Queen" dynamics of the virus against the host immune system during a single untreated infection lasting about a decade has been studied in great detail at the phenotypic level.[60] Studies of clonal distributions at the sequence level also indicate that rapid selective sweeps persist during the chronic phase (Fig. 2). Furthermore, the phylogenetic trees of random samples look remarkably like the population-level influenza evolution over decades (see e.g., Fig. 7 of Lee *et al.*[60]). The implications of these fast processes for the slow adaptive journey of the HIV virus in its host are a current subject of study.

Information hierarchies, architecture, and constraints

Biological systems utilize a variety of mechanisms at various length and time scales to store, interpret, and use information. As discussed throughout this report, the information itself is organized in a complex hierarchy: from DNA sequences, to chromatin, to tissue/organ organization, to the interactions between organisms and between species. These informational hierarchies yield layered architectures that constrain evolutionary processes at multiple levels of organization. Presentations and discussions throughout the workshop explored the informational architecture of living systems and classes of functions and architectures ranging from bacterial genomes, to the human brain and the Internet, to whether identifying general principles for structuring informational hierarchies in living systems could provide insights into the emergence of life itself.

Formation of neural networks: nature versus nurture

Alexei Koulakov (Cold Spring Harbor Laboratory) addressed the roles of nature versus nurture in the development of neural systems. Koulakov opened with a discussion of the possible role of cooking in the development of large brains in humans.[61] Brains are metabolically expensive: brain mass scales with body mass to the ¾ power (and hence also scales with metabolic rate via Kleiber's Law for metabolic scaling). This fostered a lively discussion and debate among the workshop participants about the role of cooking as a technological way of externalizing part of the digestive track. The heavy metabolic load necessary to maintain a large brain prompted Koulakov to pose the question: why have a brain?

One possible reason, as described by Koulakov, is explained by the expensive genome hypothesis, which suggests that the brain (or nervous system) is an evolutionary way to externalize parts of the adaptation process (with respect to the genome). Support for this viewpoint derives from comparing the amount of data stored in the human genome with its 3×10^9 base pairs—approximately 1 GB of data, assuming that DNA is just a linear string of bits—to the data storage capacity of the brain. Koulakov argued that the human brain, by contrast to the human genome, could store upwards of potentially 600 TB of data. This spurred the dilemma of how the approximately 1 GB of genomic information taken for a linear sequence could establish the 600 TB of information in neural network connections. Koulakov suggested that the dilemma is only resolved if synapses are not specified individually by the genome. As such, the lower-level data storage of the genome should specify a process rather than the final state of synaptic connections within the brain. He then concluded with a discussion of how models akin to those used in condensed-matter physics might be used to describe the formation of neural networks without recourse to identifying the underlying genetic mechanisms, thereby suggesting that higher-level information processing that goes beyond the flat depiction of the genome as a string of letters, as discussed throughout this report, may play an important role in the development of neural networks.

Architecture: constraints that deconstrain

John Doyle (Caltech) introduced insights into how universal laws and architecture (modularity and protocols) constrain the function and evolution of systems ranging from the biological to the technological. Several key concepts discussed by Doyle included nonconvex optimization, layered architectures as "constraints that deconstrain", and hard lim-

its ("universal laws") on robustness and efficiency. Doyle described nonconvex optimization as a feature of robust systems given that they are large (high-dimensional) but thin (even more highly codimensional) and nonconvex in the space of all possible systems.[62] This concept is analogous to the set of words in most languages, which is both large and vanishingly thin as a fraction of all possible sequences of letters. Doyle discussed how these ubiquitous features of robust systems suggest that the path toward higher levels of organization is through protocols. Doyle explained how robust architectures are constrained by protocols, but the resulting crypticity and modularity that these constraints enable also deconstrain systems designed using such architectures, enabling them to perform more diverse tasks. What emerges from this perspective is a view of architecture as a set of constraints that enables greater flexibility by increasing the accessibility of useful alternatives.

Doyle cited an interesting consequence of "universal laws" (constraints) that emerge from layered architectures: many, like the virtualization of operating systems in modern computers, are independent of the underlying physics (e.g., hardware) and therefore result in "undecidability" rather than being predictable outcomes of the underlying physical law. Such virtualization is readily apparent in biology; however, the situation in living systems is much more complex than that for technology. The example Doyle provided was of an *E. coli* cell, which is running an incredible number of "applications" relative to more familiar operating systems such as Android or iOS. The internet was also described in juxtaposition to biology in terms of robust design. Doyle noted that a key vulnerability of the internet is that TCP/IP is not layered strongly enough, lacking modern naming and virtual addressing, thus making it vulnerable to attack.[63]

The discussion then moved to the role of hard-limits, or universal laws in robust efficiency. Doyle presented glycolysis as case study demonstrating the hard trade-offs between robustness and efficiency.[64] This led to a discussion on how much of biology relies on robust control, where most genes code for control and regulatory function. Control mechanisms are put in place to manage different resources at each layer in a given architecture, where cross-layer interactions do not occur without a programmable interface. Doyle noted that conservation requires maintaining protocols, but once higher layers are added, flexibility allows lower levels to change. He concluded by suggesting that this might imply a conservation of change principle whereby some components of layered architectures are frozen to permit change of others. This could potentially have fascinating consequences for our understanding of evolutionary processes in biology.

The information hierarchy and the emergence of life

Sara Imari Walker (Arizona State University) discussed how insights from the organizational and hierarchical structure of biological systems provide new approaches to understanding the emergence of life. She began with some background on traditional approaches to the origin of life, including a survey of both genetics-first and metabolism-first perspectives. She then discussed how the standard criterion for judging the validity of these scenarios is the capacity to undergo Darwinian evolution, leading many to favor genetics-first perspectives under the current paradigm. She explained that while Darwinian evolution might be necessary to life, it isn't a sufficient condition for defining the transition from non-living to living systems. Walker instead highlighted the key role played by *information* in the logical reorganization of systems that make the transition to the living state. She suggested a framework for defining the origin of life as a transition in how information is managed and processed, corresponding to a transition in the causal flow of information within a physical system.[65]

As an illustrative example, Walker cited the early work of John von Neumann on self-reproducing machines and the distinction that must be drawn between trivial and non-trivial self-replication. She discussed how trivial self-replicators are systems that strictly rely on the implicit physics and chemistry of their host environment to support replication. Examples include memes, crystals, lipid composomes, and the non-enzymatic template–directed replication of nucleic acids. She contrasted these systems with non-trivial self-replicators such as living systems, which are distinct from trivial replicating systems in that they have some autonomy from their host environment. She described how this feature is characteristic of non-trivial self-replicators because they implement the active use of information via the readout of coded instructions to operate. Thus

trivial and non-trivial replicators differ fundamentally in the way information is organized and how it flows through the system. Examples of non-trivial systems include a von Neumann self-reproducing machine and all known life. An important note is that both trivial (e.g., memes and non-enzymatic template replicators) and non-trivial self-replicators (e.g., living organisms) can be capable of Darwinian evolution, thus evolutionary capacity does not draw a dividing line between the two classes of systems. Walker suggested that the trivial/non-trivial distinction might provide a more rigorous criterion for defining the transition from non-life to life than the Darwinian one because it relies on identifying the presence of informational hierarchies, which may be more universally characteristic of life. Discussion among the group led to the intriguing implication that the transition from non-life to life might be undecidable in the logical sense due to the strong parallels drawn between living systems and universal computing machines.

Digging deeper into the nature of this distinction, Walker discussed how the logical and organizational structure of non-trivial replicators suggests that life might be uniquely characterized by informational architecture. In this framework, candidate measures for the transition from non-life to life would rely on the causal efficacy of distributed control, which could be characterized by top-down information flow from higher to lower levels of organization in biological informational hierarchies.[66] Walker concluded by discussing how further development of this approach might lead to novel insights and approaches into understanding the emergence of life that go beyond the specific chemical substrate of life as we now know it.

Concluding remarks

The workshop brought together leading scientists exploring the structure, function, and evolution of biological systems at various length and time scales. While perhaps not yet reaching the legal standard of a "preponderance of the evidence", the theory and experiments discussed at the workshop clearly suggest that evolution is empowered by organismal architecture at multiple scales, ranging from DNA sequences and chromatin regulation all the way to interactions between organisms and species. It is not sufficient to characterize evolution as a passive selective filter of random flips along a string of letters: evolution is a much more dynamic process intimately intertwined with the constraints imposed by the informational architecture of biological systems and the interaction between an organism and its host environment. This view presents new challenges but also new opportunities, as discussed throughout this report. In particular, identifying how evolvability depends on the level of organization, versus being a scale-independent architectural feature, remains a challenging open question to be addressed. The challenge moving forward will be to develop new frameworks that can integrate the multiple scales of organization discussed throughout the workshop into quantitative predictions about the evolutionary process.

A highlight of the workshop was a lunchtime visit by a local baby bear at the Aspen Center for Physics. This visit was much to the delight of the workshop participants who were out enjoying their lunch in the fresh Colorado air, and to the dismay of those who missed the opportunity! However, a larger part of the excitement during the workshop was generated by the coming together of scientists from so many different disciplines, each bringing their own unique perspective on the interaction between evolution and information hierarchies in biological systems. The discussions generated by these interdisciplinary and cross-scale conversations provided fresh insights into many open questions surrounding organismal storage, use, and interpretation of information at multiple scales and the resultant impact on adaption. By sharing these discussions and open questions here, we hope that this report will further advance new approaches into this challenging subject by fostering similar cross-disciplinary discussion in the wider scientific community.

Acknowledgements

The Evolutionary Dynamics and Information Hierarchies workshop was held on August 19–September 9, 2012 at the Aspen Center for Physics in Aspen, CO. The authors wish to thank the hospitality of the Aspen Center for Physics for hosting the workshop, which was supported in part by the National Science Foundation under Grant no. PHY-1066293. The workshop was organized by Gyan Bhanot (Rutgers University), Lynn Caporale (St. John's University), Sebastian Doniach (Stanford University), Alexandre Morozov (Rutgers University), and Matteo Pellegrini (University of California, Los Angeles).

Conflicts of Interest

The authors declare no conflicts of interest.

References

1. Caporale, L.H. 2003. Natural selection and the emergence of a mutation phenotype. *Annu. Rev. Microbiol.* **57:** 467–485.
2. Caporale, L.H. 2006. An Overview of The Implicit Genome. In L. Caporale (Ed.), *The Implicit Genome.*, Oxford University Press. New York
3. Palmer, M., M. Lipsitch, E. Moxon & C. Bayliss. 2013. Broad conditions favor the evolution of phase-variable Loci. *mBio.* 4(1): e00430–12.
4. Doyle, J., M. Csete & L. Caporale. 2006. An Engineering Perspective: The Implicit Protocols. In L. Caporale (Ed.), *The Implicit Genome.* Oxford University Press. New York.
5. Ponder, R., N. Fonville, & S. Rosenberg. 2005. A switch from high-fidelity to error-prone DNA double-strand break repair underlies stress-induced mutation. *Mol. Cell.* **19:** 791–804.
6. Shee, C., J. Gibson, M. Darrow, C. Gonzalez & S. Rosenberg. 2011. Impact of stress-inducible switch to mutagenic repair of DNA breaks on mutation in Escherichia coli. *Proc. Natl. Acad. Sci. USA.* **108:** 13659.
7. Harris, R.S., S. Longerich & S.M. Rosenberg. 1994. Recombination in adaptive mutation. *Science* **8:** 258–260.
8. Strathern, J.N., B.K. Shafer & C.B. McGill. 1995. *Genetics* **140:** 965–972.
9. Nik-Zainal, S. *et al.* 2012. Mutational Processes Molding the Genomes of 21 Breast Cancers. *Cell.* **149:** 979–993.
10. Roberts, S.A. *et al.* 2012. Clustered Mutations in Yeast and in Human Cancers Can Arise from Damaged Long Single-Strand DNA Regions. *Mol. Cell* **46:** 424–435.
11. Shee, C., J. Gibson, & S. Rosenberg. 2012. Two Mechanisms Produce Mutation Hotspots at DNA Breaks in Escherichia Coli. *Cell Rep.* **2:** 714–721.
12. Al Mamun, A.A.M. *et al.* 2012. Identity and Function of a Large Gene Network Underlying Mutagenic Repair of DNA Breaks. *Science* **338:** 1344–1348.
13. Schuster-Bockler, B. & B. Lehner. 2012. Chromatin organization is a major influence onf regional mutation rates in human cancer cells. *Nature* **488:** 504–507.
14. Dobrovolskaia, I. & G. Arya. 2012. Dynamics of forced nucleosome unraveling and the role of nonuniform histone-DNA interactions. *Biophys. J.* **103:** 989–998.
15. Grigoryev, S., G. Arya C., Correll, *et al.* 2009. Evidence for heteromorphic chromatin fibers from analysis of nucleosome interactions. *Proc. Natl. Acad. Sci. USA* **106:** 13317–13322.
16. Arya, G. & T. Schlick. 2009. A tale of tails: how histone tails mediate chromatin compaction in different salt and linker histone environments. *J. Phy. Chem. A* **113:** 4045–4059.
17. Meluzzi, D. & G. Arya. 2013. Recovering ensembles of chromatin conformations from contact probabilities. *Nucleic Acids Res.* **41:** 63–75.

18. Correll, S., M.H. Schubert, & S. Grigoryev. 2012. Short nucleosome repeats impose rotational modulation on chromatin fibre folding. *EMBO J.* **31:** 2416–2426.

19. Dodd, I., M. Micheelsen, K. Sneppen & G. Thon. 2007. Theoretical analysis of epigenetic cell memory by nucleosome modification. *Cell* **129:** 813–822.

20. Angel, A., J. Song, C. Dean & M. Howard. 2011. A Polycomb-based switch underlying quantitative epigenetic memory. *Nature* **476:** 105–108.

21. Probst, A.V. & G. Almouzni. 2008. Pericentric heterochromatin: dynamic organization during early development in mammals. *Differentiation* **76:** 15–23.

22. Müller, K.P. *et al.* 2009. Multiscale analysis of dynamics and interactions of heterochromatin protein 1 by fluorescence fluctuation microscopy. *Biophys. J.* **97:** 2876–2885.

23. Erdel, F., K. Müller-Ott, M. Baum, *et al.* 2011. Dissecting chromatin interactions in living cells from protein mobility maps. *Chromosome Res.* **19:** 99–115.

24. Rippe, K. 2001. Making contacts on a nucleic acid polymer. *Trends Biochem. Sci.* **26:** 733–740.

25. Rippe, K., P.H. Hippel & J. Langowski. 1995. Action at a distance: DNA-looping and initiation of transcription. *Trends Biochem. Sci.* **20:** 500–506.

26. Ringrose, L., S. Chabanis, P.-O. Angrand, *et al.* 1999. Quantitative comparison of DNA looping in vitro and in vivo: chromatin increases effective DNA flexibility at short distances. *EMBO J.* **18:** 6630–6641.

27. Chodavarapu, R.K. *et al.* 2010. Relationship between nucleosome positioning and DNA methylation. *Nature* **466:** 388–392.

28. Schwartz, S., E. Meshorer & G. Ast. 2009. Chromatin organization marks exon-intron structure. *Nat. Struct. Mol. Biol.* **16:** 990–995.

29. Berger, S.L. 2007. The complex language of chromatin regulation during transcription. *Nature* **447:** 407–412.

30. Werken, H.J. *et al.* 2012. Robust 4C-seq data analysis to screen for regulatory DNA interactions. *Nat. Methods* **9:** 969–972.

31. Powell, S., D. Szklarczyk, K. Trachana, *et al.* 2012. eggNOG v3.0: orthologous groups covering 1133 organisms at 41 different taxonomic ranges. *Nucleic Acids. Res.* **40:** D284–D289.

32. Pang, T. & S. Maslov. 2011. A Toolbox Model of Evolution of Metabolic Pathways on Networks of Arbitrary Topology. *PLoS Comput. Biol.* **7:** e1001137.

33. Nash, T. 2008. *Lichen Biology.* Cambridge University Press. Cambridge.

34. Mathiesen, J., N. Mitarai, K. Sneppen, & A. Trusina. 2011. Ecosystems with Mutually Exclusive Interactions Self-Organize to a State of High Diversity. *Phys. Rev. Lett.* **107:** 188101.

35. Mitarai, N., J. Mathiesen & K. Sneppen. 2012. Emergence of diversity in a model ecosystem. *Phys. Rev. E* **86:** 011929.

36. Schwede, A. & M. Carrington. 2010. Bloodstream form Trypanosome plasma membrane proteins: antigenic variation and invariant antigens. *Parasitology* **137:** 2029–2039.

37. Barry, J. & R. McCulloch. 2001. Antigenic variation in trypanosomes: enhanced phenotypic variation in a eukaryotic parasite. *Adv. Parasitol.* **49:** 1–70.

38. Barry, J. D., J.P. Hall & L. Plenderleith. 2012. Genome hyperevolution and the success of a parasite. *Ann. N.Y. Acad. Sci.* **1267:** 11–17.

39. Berriman, M. *et al.* 2005. The genome of the African trypanosome Trypanosoma brucei. *Science* **309:** 416–422.

40. Marcello, L. & J. Barry. 2007. Analysis of the VSG gene silent archive in Trypanosoma brucei reveals that mosaic gene expression is prominent in antigenic variation and is favored by archive substructure. *Genome Res.* **17:** 1344–1352.

41. Boothroyd, C.E. *et al.* 2009. A yeast-endonuclease-generated DNA break induces antigenic switching in Trypanosoma brucei. *Nature* **459:** 278–281.

42. Cliffe, L., T. Siegel, M. Marshall, *et al.* 2010. Two thymidine hydroxylases differentially regulate the formation of glucosylated DNA at regions flanking polymerase II polycitronic transcription units throughout the genome of Typanosome brucei. *Nuc. Acids. Res.* **38:** 3923–3935.

43. Tiengwe, C. *et al.* 2012. Genome-wide analysis reveals extensive functional interaction between DNA replication initiation and transcription in the genome of Trypanosome bruce. *Cell Rep.* **2:** 185–197.

44. Tiengwe, C. *et al.* 2012. Identification of ORC1/CDC6-interacting factors in Trypanosoma brucei reveals critical features of origin recognition complex architecture. *PLoS One* **7:** e32674.

45. Strelkowa, N. & M. Lässig. 2012. Clonal Interference in the Evolution of Influenza. *Genetics* **192:** 671–682.

46. Mustonen, V. & M. Lässig. 2010. Fitness flux and ubiquity of adaptive evolution. *Proc. Natl. Acad. Sci. USA* **107:** 4248–4253.

47. Foley, B. 2000. An overview of the molecular phylogeny of lentiviruses. In Kuiken, C. *et al.* (Ed.), *HIV Sequence Compendium 200: Theoretical Biology and Biophysics* (pp. 35–43). Los Alamos National Lab. Los Alamos.

48. Keele, B. *et al.* 2009. Increased mortality and AIDS-like immunopathology in wild chimpanzees infected with SIVcpz. *Nature* **460:** 515–519.

49. Keele, B. *et al.* 2006. Chimpanzee reservoirs of pandemic and nonpandemic HIV-1. *Science* **313:** 523–526.

50. Korber, B. *et al.* 2000. Timing the Ancestor of the HIV-1 Pandemic Strains. *Science* **288:** 1789–1796.

51. Barre-Sinoussi, F. *et al.* 1983. Isolation of a T-lymphotropic retrovirus from a patient at risk for acquired immune deficiency syndrome (AIDS). *Science* **220:** 868–871.

52. Gallo, R. *et al.* 1984. Frequent detection and isolation of cytopathic retroviruses (HTLV-III) from patients with AIDS and at risk for AIDS. *Science* **224:** 500–502.

53. Yusim, K. *et al.* 2001. Using human immunodeficiency virus type 1 sequences to infer historical features of the acquired immune deficiency syndrome epidemic and human immunodeficiency virus evolution. *Proc. Roy. Soc. B* **356:** 855–866.

54. Mansky, L. & H. Temin. 1995. Lower in vivo mutation rate of human immunodeficiency virus type 1 than that predicted from the fidelity of purified reverse transcriptase. *J. Virol.* **69**(8): 5087–5094.

55. Stafford, M., L. Corey, Y. Cao, *et al.* 2000. Modeling Plasma Virus Concentration during Primary HIV Infection. *J. Theo. Biol.* **203:** 285–301.

56. Moore, C., M. John, I. James, *et al.* 2002. Evidence of HIV-1 Adaptation to HLA-Restricted Immune Responses at a Population Level. *Science* **296:** 1439.

57. Bhattacharya, T. *et al.* 2007. Founder Effects in the Assessment of HIV Polymorphisms and HLA Allele Associations. *Science* **315:** 1583–1586.

58. Fischer, W. *et al.* 2010. Transmission of Single HIV-1 Genomes and Dynamics of Early Immune Escape Revealed by Ultra-Deep Sequencing. *PLoS One* **5:** e1000890.

59. Richman, D., T. Wrin, S. Little & C. Petropoulos. 2003. Rapid evolution of the neutralizing antibody response to HIV type 1 infection. *Proc. Natl. Acad. Sci. USA* **100:** 4144–4149.

60. Lee, H., A. Perelson, S.-C. Park & T. Leitner. 2008. Dynamic Correlation between Intrahost HIV-1 Quasispecies Evolution and Disease Progression. *PLoS Compt. Biol.* **4:** e1000240.

61. Aiello, L. & P. Wheeler. 1995. The Expensive-Tissue Hypothesis: The Brain and the Digestive System in Human and Primate Evolution. *Current Anthropology* **36:** 199–221.

62. Doyle, J. & M. Csete. 2011. Architecture, constraints, and behavior. *Proc. Natl. Acad. Sci. USA* **108:** 15624.

63. Doyle, J. *et al.* 2005. The "robust yet fragile" nature of the Internet. *Proc. Natl. Acad. Sci. USA* **102:** 14497.

64. Chandra, F., G. Buzi & J. Doyle. 2011. Glycolytic oscillations and limits on robust efficiency. *Science* **333:** 187–192.

65. Walker, S.I. & P. Davies. 2013. The algorithmic origins of life. *J. Roy. Soc. Interface* **10:** 20120869.

66. Walker, S.I., L. Cisneros & P. Davies. 2012. Evolutionary transitions and top-down causation. *Proceedings of Artificial Life XIII* (pp. 283–290). MIT Press. Cambridge.

Ann. N.Y. Acad. Sci. ISSN 0077-8923

ANNALS OF THE NEW YORK ACADEMY OF SCIENCES

Issue: *Evolutionary Dynamics and Information Hierarchies in Biological Systems*

In Darwinian evolution, feedback from natural selection leads to biased mutations

Lynn Helena Caporale and John Doyle[1]

[1]Control and Dynamical Systems California Institute of Technology, Pasadena, California 91125

Address for correspondence: Lynn Helena Caporale 1 Sherman Square, New York, NY 10023. caporale@usa.net

Natural selection provides feedback through which information about the environment and its recurring challenges is captured, inherited, and accumulated within genomes in the form of variations that contribute to survival. The variation upon which natural selection acts is generally described as "random." Yet evidence has been mounting for decades, from such phenomena as mutation hotspots, horizontal gene transfer, and highly mutable repetitive sequences, that variation is far from the simplifying idealization of random processes as *white* (uniform in space and time and independent of the environment or context). This paper focuses on what is known about the generation and control of mutational variation, emphasizing that it is not uniform across the genome or in time, not unstructured with respect to survival, and is neither memoryless nor independent of the (also far from white) environment. We suggest that, as opposed to frequentist methods, Bayesian analysis could capture the evolution of nonuniform probabilities of distinct classes of mutation, and argue not only that the locations, styles, and timing of real mutations are not correctly modeled as generated by a white noise random process, but that such a process would be inconsistent with evolutionary theory.

Keywords: Darwin; random mutation; natural selection; evolution; feedback

Introduction

While the term *random mutation* is widely used as if it were the original foundation of evolutionary theory, Darwin explicitly stated:

> I have . . . sometimes spoken as if the variations . . . had been due to chance. This, of course, is a wholly incorrect expression, but it serves to acknowledge plainly our ignorance of the cause of each particular variation.[1]

A century before biochemists began to describe the underpinnings of variation among individuals, it was attention to variation that led to the theory of evolution. Charles Darwin and Alfred Russel Wallace were perceptive naturalists and collectors, who noticed variation among individuals of the same species.[a] They proposed that those variants that are most fitted to their environment pass on traits that contributed to fitness in the next generation, leading to descent with modification and adaptation,[2,3] although they did not know how traits were modified or inherited.

With respect to Mendel's observations,[4] Wallace wrote:

> The essential basis of evolution, involving as it does the most minute and all-pervading adaptation to the whole environment, is *extreme and ever-present plasticity, as a condition of survival and adaptation.*[b] But the essence of Mendelian characters is their rigidity. They are transmitted without variation, and therefore, except by the rarest of accidents, can never become adapted to ever varying conditions.[5]

name "this principle, by which each slight variation is selected, Natural Selection."[2]

---

[a]"considering the amount of individual variation that . . . experience as a collector had shown . . . to exist . . . ;"[3]

[b]Emphasis added.

doi: 10.1111/nyas.12235

Mendel's observations were integrated into evolutionary theory[6] through the concept that variation within populations results from different versions (alleles) of genes (Mendel's inherited characters). The statistician Ronald Fisher[6,7] introduced a geometric model with the mathematical assumption that generation of the phenotypic manifestations of variation could be represented as a normally distributed random process. What was implicit in this model is that mutation creates new alleles with white noise increments and then selection alters the prevalence of different alleles. Thus, the concept of *random* was attached to evolutionary theory not by Darwin but over half a century later.

That mutations are random is taught now as an integral part of evolutionary theory. For example, a book targeted to high school teachers, filled with engaging examples of ways to teach evolution, describes "Darwin's law of natural selection" as stating that "descent with modification and adaptation result from the natural selection of heritable random variations."[8] The term *law* is used instead of *theory* due to misunderstanding by those not trained in science about the meaning of *theory*,[d] but, there is at least as much confusion about the meaning of *random*.

There has been a narrowing within physics and science generally of the term *random* to be nearly synonymous with *white noise*. An assumption that variation is a random white noise process would mean that mutations are uniform (or unbiased) with respect to position along the chromosome, stationary (i.e., uniform also with respect to time), memoryless (independent of past changes), and independent of context or environment (i.e., autonomous). These assumptions simplify modeling and analysis because mutations thus restricted

would render genome sequences a random walk with white noise increments, and selection simply would favor those fittest to the environment. However, for real biology to preserve this idealization would require selection to have no feedback that could act on the mechanisms that generate variation, as this would likely disrupt its white noise properties.

In the rest of this paper, we will discuss how well-known experiments clearly show that real biology systematically and routinely violates all of the white noise assumptions. We will emphasize that feedback from selection to the biochemical processes that generate variation would make mutation very far from white by increasing the probability that variants will survive (compared to variants generated by a random white noise process). We then briefly end with a discussion of the implications for a more rigorous mathematical formalization of evolutionary theory, which will be pursued in subsequent papers.

Mutation is not uniform with respect to position along the DNA

The term *random mutation* was attached to evolutionary theory before the chemical nature of mutation was understood. Even so, the assumption of uniformity (unbiased) with respect to position along the chromosome was known to be violated[9,10] even before the Avery laboratory demonstrated[11] that the chemical underlying transmission of inherited characters is DNA. Mutation hotspots were observed almost as soon as DNA was understood to be the genetic material; as Seymour Benzer commented in 1961 regarding the genetic map of the phage T4 that "the distribution is nonrandom leaps to the eye."[12] The availability of genome sequences has made the nonuniform distribution of mutations noted by Benzer all the more obvious.[e,13,14]

Even without this evidence, a thoughtful biochemist can predict that it would be highly unlikely for mutations to be uniformly distributed along a DNA sequence. Although routinely treated as text in search algorithms, DNA bases are, of course, not actually letters. DNA double helices are physical chemical entities with distinct properties,[15] such

---

[c] "The possible positions representing adaptations superior to that represented by A will be enclosed by a sphere passing through A and centred at O. If A is shifted through a fixed distance, r, *in any direction* its translation will improve the adaptation if it is carried to a point within this sphere, but will impair it if the new position is outside." (emphasis added; note that in his discussion shifts can represent changes either in the organism's phenotype or its environment).

[d] As in, "oh, it's just a theory."

[e] "It is clear that the mammalian genome is evolving under the influence of non-uniform local forces."[13]

as the uneven tilt and twist of base pairs (which cause sequence-dependent deviations from the iconic DNA structure),[16] repetitive sequences that tend to misalign and slip, and many noncanonical structures.[17–19] Such sequence context–dependent variations in physical chemical properties result in often dramatic sequence context–dependent effects on the fidelity of the enzymes that repeatedly copy and repair DNA, which in turn affects the evolution of that DNA sequence.

Thus, even if one were to synthesize a computer-generated white noise sequence of nucleotides, when biochemically replicated it would reveal intrinsic sequence context–dependent variations in the probability of distinct classes of mutation. In other words, the probability of distinct classes of mutation would not be expected to be random white noise, and thus not be uniformly distributed along the DNA molecule (or chromosome). This simple description of biochemistry has implications for evolutionary theory, as described in the next section.

Generation of variation can be biased as to whether a mutation will contribute to fitness

Since the early 20th century, it has been argued that most mutations must be deleterious, although these discussions, which began prior to an understanding of the biochemistry of mutation, did not consider local contexts that can affect the probability of distinct classes of mutation; further, as referenced in a discussion by David King, "for mutations arising spontaneously under natural conditions," (i.e., in contrast to mutations created in the laboratory) "the ratio of benefit to harm has never been realistically assessed."[20]

In contrast to our assertion that feedback from natural selection would affect the probability of distinct classes of mutation, it has been argued that natural selection cannot "assist the process of evolutionary change" as "selection lacks foresight and no one has described a plausible way to provide it."[21] Thus, in statements of evolutionary theory, the assertion is made that mutation is random with respect to its probability of being adaptive.[f] However, many classes of environmental challenge recur. Hosts combat pathogens (and pathogens avoid

host defenses); predators and prey do battle through biochemical adaptations;[22] bird beaks must pick up and crack[23] available seeds (or insects)—a menu that may change rapidly due, for example, to a drought.

It is important to emphasize that not all random processes are well modeled as white noise. Thus we can ask whether there are biases in random mutation. There are of course rare environmental events, such as meteor strikes, and a wide range of processes that cause mutation with low probability at any site (it is to these events that the historical objections reviewed in Ref. 20 are likely to apply). However, environments change in ways that have structure, which would select for variation that has compensatory structure. For example, if a pathogen's environment contains a host immune system that continually generates new antibodies directed specifically against the pathogen's coat, mechanisms that generate rapid, focused, but still probabilistic variation in coats would be expected to be favored by selection.[24] Of course, the extent to which selection acts at such sites depends on the challenge/opportunity.[25,26]

In fact, since the probability of mutation varies along the DNA, and since the theory of evolution states that selection acts on variation, the theory of evolution actually predicts that variation will become at least somewhat structured, with mutations focused through mechanisms that could not have been taken into account before the biochemistry of mutations was investigated.[27] Selection should act on the biochemistry of genome variation much as it acts on beaks and wings. In other words, due to the repeated action of selection, an assertion that mutation is uniform, memoryless, and stationary, far from being integral to Darwin's insights, would be an implausible idealization that is inconsistent with evolutionary theory.

But in what sense could DNA sequences incorporate the potential to generate variants with an increased likelihood of surviving another round of selection? As more examples are found, our ability to imagine expands. For example selection can deplete mutable sequences,[g] such as repeats that tend to misalign and gain and lose units, from regions where variation is harmful, but such mutable

---

[f]"Mutation is a random process with respect to the adaptive needs of the species"—T. Dobzhansky.[6]

[g]By selecting against organisms that inherit mutable sequences at loci where they damage essential functions at a high rate.

sequences are enriched[20,28] in proteins involved in interactions with a changing environment. For example, in proteins that affect a pathogen's coat, loss or gain of a unit of a tetrameric repeat (such as CAATCAATCAATCAAT) shifts the reading frame, leading to loss of a coat variant recognized by the immune system (and which thus would have targeted the individual for destruction), at rates that are orders of magnitude above that of the background mutation rate (i.e., genome-wide average nucleotide substitution rate).[29,30]

Gain or loss of even a single unit in repeats such as GGGGGGGGGG or GGGGGGGGGGGG has a dramatic effect on the strength of binding of transcription factors, as it rotates the position of the −10 and −35 bacterial promoter consensus sequences around the helix relative to each other.[31] In the eukaryote *Saccharomyces cerevisiae*, it has been reported that ∼25% of genes have tandem repeats in promoters, affecting expression levels.[32] Thus, there is significant standing variation in populations of individuals descended from a common ancestor bearing such mutable sequences.[20]

Note that a mutation involving loss or gain of units in repetitive sequences is reversible. Because such mutations are reversible, descendants are not trapped on a narrow fitness peak of the moment; rather, any individual has the potential to generate, among its population of descendants, a range of variants, facilitating survival of descendants as they confront the challenges and opportunities of a wider range of environments. Such highly repetitive sequences are found not only in bacteria, but also in eukaryotes,[33,34] including people.[35]

In contrast to recessive alleles, which are explicitly present in the genome, alternative genotypes that arise from slips in repetitive DNA sequences can be viewed as implied by the sequence that encodes them. Nonallelic diversity also is implied by sequences that raise the probability of gene duplication.[36] Thus, the probability of distinct classes of mutations has been biased, making some classes of mutation (such as changes in pathogen coats) more likely than others. Further, for some cases, such as a change of coat, any change may well protect the pathogen. This is in contrast to a change in an antibody-binding site, where, while generation of variation has focused on the variable region, a repertoire must be generated by targeted mutation in order to create the antibody that will bind

to the pathogen's new coat. Horizontal gene transfer (HGT) is an obvious and dramatic example of a mechanism that expedites creation of a set of accessible functional genomes that is vastly larger than would be possible with only white noise mutation.[37]

Dobzhansky wrote,[6] "only a vitalist Pangloss can imagine that the genes know how and when it is good for them to mutate." This makes sense when considering whether one isolated individual nucleotide could know whether it might be better to be an A or a G in the next generation. We now understand that each nucleotide is embedded in a context, from genomic to environmental, with correlations emerging via selection over evolutionary time scales, and that this context can affect the fate of that nucleotide.

While locations in the genome may present with different probabilities of beneficial versus deleterious changes relative to genome averages, the direction of an individual mutation nevertheless still can be viewed as essentially randomly generated, but by a process that has been biased by selection. This process generates mutations that are not accurately modeled as random white noise. Note that saying a process is not modeled well by white noise is not equivalent to saying the process is not random (i.e., not probabilistic). Note also the difference between focusing on the probability of any one individual mutation and the probability of a mutation arising in a population. For example, we can say that for one individual bacterium's DNA it is (biased) random whether the change of length of a repetitive sequence takes it in the direction that would contribute to survival; however, given the size of bacterial populations, for loci that mutate at rates as high as 1/1000 and higher,[29] the population would almost certainly contain the variant that would survive the genetically anticipated challenge.

In other words, mechanisms that generate variation can adapt to a recurring nonuniform distribution of challenges, and thus in effect have a type of memory that would generate variation that is still random (i.e., probabilistic) but with an increased probability of generating variants that survive classes of challenges the genome and its descendants are likely to face in the future (i.e., if they are the same classes of challenges that the genome's lineage survived in the past). Thus, the statement that "all mutation is random"—in the sense of unstructured and uniform—is inconsistent not only with a growing body of data, but also with the theory of evolution, due to the repeated effects of selection on mechanisms that generate mutations.

The mutation rate is neither stationary in time nor independent of the environment

While faithful reproduction of the genome transmits adaptations from generation to generation, those lineages that do not generate any diversity may be vulnerable[6] to, for example, a pathogen, or sudden loss of a food source. Thus, a balance between fidelity and exploration would be expected to evolve. That the rate of mutation does not change over time had been the assumption underlying the use of molecular clocks.[38] But is the probability of each class of mutation really unaffected by environmental or other influences that change over time?

As described in more detail below, the potential adaptive value of variation is constant neither across the genome nor over time. Thus stability (protection of adaptations) and diversity (exploration of new adaptations) can be balanced by an increased probability of variation not only in certain regions of the genome but also at times when the organism finds itself poorly adapted to its environment.

Since generation of variation results from biochemical processes, generation of variation, like biochemical processes, can be regulated. Biochemical mechanisms are available that can focus variation on different regions of the genome by, for example, induction of different sets of enzymes during different times during replication,[14,36] an effect that is increasingly accessible to analysis.[39]

This paper arises from a workshop centered on consideration of information hierarchies in biological systems.[40] Organisms sense, and respond with regulated metabolic changes to, the stress of starvation;[41] similarly, organisms sense and respond to the stress of DNA damage.[42] Thus, as sensing and signaling mechanisms are in place[43] that respond to the type and extent of the stress, biochemical mechanisms that affect genome variation, and thus affect evolution, can evolve connections downstream of signals that sense specific changes in the environment and specific classes of stress. For example, *Escherichia coli* senses and makes genetic changes in response to specific external clues. When it senses it is in the host environment, due to temperature and the presence of specific metabolites, a re-

versible mutation, inversion,[h] causes phase switching of fimbrae.[45]

The vertebrate immune system also demonstrates that the location and timing of distinct classes of genetic variation can be regulated. Targeted biochemical reactions generate variation in specific cell lineages (e.g., V/D/J rearrangement,[46] hypermutation[47]) and, in response to the environment, induce directed gene rearrangements (e.g., immunoglobulin class switch[48]). There is no reason to assume that such regulated, targeted variation would be unavailable to the germline.

Have connections between genetic variation and stress in fact been observed? *Arabidopsis* stressed by the presence of pathogens were observed to increase somatic recombination.[49] Barbara McClintock observed increased variation in response to the stress of DNA breakage and suggested that a cell is able to sense that it is under stress and that this might set in motion the orderly sequence of events that will mitigate this danger[50] and even trigger genome restructuring.[51] As Richard Jorgensen summarized,[i,52] McClintock proposed "a complex process that integrates information" and that could distinguish among, and mount appropriate distinct responses to, distinct classes of challenges.[53]

In documenting the generation of new regulatory networks and the apparent sudden burst of transposition by *mPing* in rice, under the stress of adapting to a temperate climate, Naito *et al.*[54] suggested that for selfing plants, bursts of transposable elements may generate genetic diversity rapidly, but also suggested that this is not limited to plants, as

---

[h]A vertebrate example of recurrent mutation involving inversion was revealed in a genomic study of "reuse of globally shared standing genetic variation, including chromosomal inversions, [which] has an important role in repeated evolution of distinct marine and freshwater sticklebacks,"[44] although there is no evidence yet that either addresses the question whether this inversion is induced rather than selected from standing variation or that standing variation is increased at that locus. Dobzhansky found evidence for seasonal variation of the prevalence of chromosome inversion as an adaptive trait in *Drosophila*.[6]

[i] "To paraphrase McClintock (1978), it is time to explore the nature and evolutionary significance of these attentive systems for adaptive genome restructuring in response to stress, "*the consequences of which vary according to the nature of the challenge to be met.*"

evidence for the rapid bursts of miniature inverted-repeat transposable elements is found in "virtually all sequenced eukaryotic genomes."

Increased mutation, observed in bacteria stressed by DNA damage or starvation, depends upon the activation of specific gene products.[55–58] and thus is not simply the result of inability to cope with the damage. Laboratory activation of the SOS DNA-damage and the RpoS-general/starvation stress response was sufficient to trigger a mutagenic mode of DNA break repair and thus increased mutation without an external stress.[58] In other words, the bacterium interprets induction of certain pathways as a biochemical signal that it is stressed. Having sequenced thousands of genomes, and determined expression patterns under varying conditions, it is possible to begin to examine genome wiring to explore whether and how distinct types of stress (and other environmental signals) might connect to pathways that affect distinct classes of variation of genome sequences.[59,60]

While evolution of responses to the environment that occur within the lifetime of an individual are widely acknowledged, there is no reason to limit biochemical responses to the environment to those that affect only a generation, since lineages survive over evolutionary timescales. It is important to note the role of feedback between selection and mechanisms that generate genome variation. Generators of diversity fall under selective pressure owing to the effects on survival of spatial and temporal biases of the classes of mutation that they generate.

Mutation, repair, and recombination depend upon biochemical processes, which can fall under the control of a wide range of regulatory systems. Thus, we cannot assume that mutation is stationary, unaffected by the environment, or constant in time.

What can evolve?

There is much more to understand about evolution than traits observed by naturalists in the field and base-by-base changes in DNA sequences observed in the laboratory. The genome is organized, with hierarchies of recognition and control. An evolutionary perspective is essential to comprehending this organization,[j,61] as is a perspective that includes feedback control and dynamics. In turn, a perspective built upon understanding this organized complexity[62] and its contexts will lead us to a deeper understanding of evolution. As we analyze genomic sequences, attention to structured and nonwhite forms of variation is likely to inform us of challenges that a lineage faces and that have exerted selective pressure during its evolution.

The initial step toward our ability to decipher information carried by DNA was to "crack the genetic code,"[63] but our work is not done. The degeneracy of the table of codons[64] and the existence of extensive[13] nonprotein coding regions leaves room to transmit additional messages underneath and around a protein-coding sequence, including messages that modulate the rate and type of genetic change. For example, the same amino acid sequence can be specified by either a mutable repetitive sequence or a more stable nonrepetitive sequence.[24] Thus, the same sequence can both specify a protein sequence and structure variation by implying mutability (i.e., affect the probability that descendants will be diverse at specific places in the genome). Such intertwined information represents an efficient use of genomic space.

Considering genome organization, self-reference, and behavior, an informative way to describe what can evolve is suggested by applying the term *protocol*[k,65] to genomics, a term used in engineering for sets of rules by which components interact to create new levels of functionality (familiar for enabling transmission of information through the internet). The table of codons is a familiar example of a protocol in biology (shared codons is one of many shared protocols necessary to enable HGT).[66] Protocols provide a useful concept for discussing evolution, including evolution of the genetic code,[67] as well as the labeled fragments with rules for their assembly that structure generation both of somatic diversity in the vertebrate immune system[46] and of the diverse repertoire of trypanosome coat proteins.[68] There are qualitatively distinct forms of information in genomes that may be nonlinear and dependent upon genomic context and relationships among sequences (such as inverted repeats or more complex structures[69]).[14]

---

[j] "Biological organization will never be understood except as the expression of an underlying evolutionary process."—Woese and Goldenfeld.[61]

[k] From the Greek use of *protocollon*, which referred to a leaf of paper glued to and labeling a manuscript scroll, defining its contents http://www.linfo.org/protocol.html

The profound importance of context, and an astounding sense of the complexity of organization that defines different contexts and behaviors of the genome, jumps to our attention in the formation of the macronucleus in ciliates, with genome-wide and predictable silencing, DNA deletions, inversions, and amplifications, built upon recognition and regulation involving RNA.[70–72]

In evolution, there would be a selective advantage for descendants of an individual that evolved an active framework that focuses exploration, compared to descendants of individuals that have a uniform probability of trying every mutation and every insertion site. This suggests the possibility that the DNA sequence of large gene families may represent a successful evolutionary framework, much as the protein sequence represents a successful functional framework.[73]

The biochemical infrastructure[74] that enables HGT in bacteria enormously increases (over white noise) the probability the bacteria will gain access to life-saving information, compared to if they lacked such infrastructure. Recent sequencing of roughly 100 *E. coli* strains (including subspecies Shigella spp) found that the genes universally shared, the median number per strain, and the total across all strains were approximately 1000, 4000, and 20,000, with the first and last numbers expected to continue to diverge as more strains are sequenced.[75] Further, in the canonical example of acquisition of antibiotic resistance by HGT, bacterial survivors in an antibiotic-rich environment (e.g., a hospital) would be a rich source for sensitive bacteria to tap for horizontal acquisition of genes conferring antibiotic resistance (the environment has structure with respect to the availability of genes accessible through HGT such that, e.g., genes encoding antibiotic resistance are most likely to be accessible from a neighboring bacterium just when and where an antibiotic-sensitive bacterium needs them).

Similarly, for eukaryotic parasites, the ability to vary coats rapidly through site-directed recombination provides a selective advantage compared to a probability of either uniformly distributed base changes or of recombination uniformly distributed in its genome.[68] Another eukaryotic example[20] of an important framework that facilitates generation of nonwhite diversity and exploration of new adap-

tations, to which significant resources are devoted, is meiosis.

Natural selection has embedded innate knowledge about the world within surviving genomes, embodied in diverse mechanisms. There are deep theories in systems engineering that help explain not just how such mechanisms that incorporate information about the environment work, but why they are necessary for robust performance. A richer evolutionary theory could incorporate analysis of how regularities in the environment's dynamics can become embedded in genomes in the form of dynamics of control circuits, but this has so far been explored in only a few settings.[76,77] We are very familiar with control circuits responsive to regularities of the environment that operate within a generation, such as circadian rhythms.[78] In another familiar example, using innate circuitry devoted to this purpose, *E. coli* swim toward glucose.[79] This circuitry obviously embodies structural models of attractants in the receptor proteins that recognize them, but the dynamics of adaptation circuitry also embodies a model of the structure of the environment (i.e., a model of the direction in which the concentration of the attractant increases). Furthermore, the necessity of these internal models can be made mathematically precise.[77]

The sophistication of internal models appears to increase with greater organism complexity. *Caenorhabditis elegans* inherit both innate chemosensory attractant and avoidance behaviors and mechanisms for individuals to adapt based upon experience.[80] Human sensorimotor control circuits, from basic reflexes to the most sophisticated learned skills, involve mechanisms that vary greatly in speed and flexibility, but all depend on internal models of the dynamics of the body, its extension via tools, and the environment.[81] Thus, internal models of regularities in the dynamics of its environment can be expected to contribute to an organism's fitness. Such models of the environment can also become embedded in the mechanisms that generate variation across generations. For example, innate information that structures the generation of variation underlies genomic mechanisms that facilitate our ability to create an antibody directed against an antigen never previously encountered by an individual or ancestors.[82]

Treating mutations as hypotheses about survival in an environment

Wallace rejected the predictable reassortment of the characters Mendel observed as a mechanism of evolution and instead saw that Mendel's work spoke to the stability of the inheritance of information from generation to generation. In fact, careful in his experimental design, and seeking mathematical laws, Mendel chose the characters he studied in his well-tended pea plants with care, as true breeding with two clearly distinguishable forms.[4] How startled might he have been if, confronted with Barbara McClintock's maize that had been stressed by DNA damage, he had observed the suddenly spotted kernels![53]

Much as the concept of genes as independently assorting fixed units of inheritance was shaped by Mendel's attention to an experimental design focused on stable, easily distinguished characters, evolutionary theory was shaped by statisticians whose work emphasized variance around a mean and random sampling. Sharon Bertsch McGrayne contrasts Fisher's approach to that of Bayesian statistics in her summary of Turing's words[83] regarding Enigma cryptanalysis: "confirming inferences suggested by a hypothesis would make the hypothesis itself more probable."[84] How might a Bayesian perspective be applied to evolutionary theory? Distinct classes of mutation could be modeled differently,[85] but beyond this, suppose the set of mutations (or lack of them) generated in each individual's gametes was treated as a prior model or hypothesis about survival: then survival of descendants is an observation. Assumptions of the model (including the assumptions underlying various understandings of random mutation outlined above) would be tested over many generations, with the model updated based upon observations (i.e., the selection and survival of descendants bearing variation generated along the genome by diverse processes).

Such a Bayesian view would predict that evolution itself would drive genomes away from white noise variations, not merely to nonuniform mutations, but ultimately toward embodied models of environmental regularities. The above catalog of mechanisms more than hints that this is possible. Darwin proposed what now looks like a feedback control engineering theory of evolution, but subsequent formalizations have interpreted it in terms of information theory and statistical physics, with minimal feedback. But variation and selection together represent essential elements in a feedback loop, and variation is not outside that loop, however appealingly simple such an assumption may be. Thus, an essential concept is that feedback from the environment, operating through (selection of) surviving descendants, must inevitably incorporate a worldview into the mechanisms that generate genome variation and genome contexts and DNA sequences that encode them, such that the probability of distinct classes of mutation can become aligned with probable effects on survival.

On reflection, Darwin appears to have been reaching for this concept when he suggested that, "deviations of structure are in some way due to the nature of the conditions of life, to which the parents and their more remote ancestors have been exposed during several generations."[2]

Wallace and Darwin began with attention to variation. Now, we are in a position to focus on the organization and regulation of biochemical mechanisms underlying generation of that variation. In fact, Darwin recognized that "a grand and almost untrodden field of inquiry will be opened, on the causes and laws of variation."[2]

Natural selection has led to the evolution, in genomes, of information that structures exploration and facilitates successful adaptation to likely challenges; thus, the most revealing and intriguing aspect of mutation, the generation of variation, is not that it is random, but rather the ways and extent to which it may become biased by feedback from selection.

Acknowledgments

The authors are deeply indebted to David King for his important and insightful comments on an earlier version of this manuscript, and recommend his recent article, which contributes historical background for this discussion.[20] This work was supported in part by the National Science Foundation under Grant No. PHYS-1066293 and the hospitality of the Aspen Center for Physics. J.D. was in part supported by the NSF, AFOSR, and the Institute for Collaborative Biotechnologies through Grant W911NF-09-0001 from the U.S. Army Research Office. The content does not necessarily reflect the position or the policy of the Government, and no official endorsement should be inferred.

Conflicts of interest

The authors declare no conflicts of interest.

References

1. Darwin, C. 1859. On the origin of the species. http://www.talkorigins.org/faqs/origin.html Chapter 5: Laws of Variation.
2. Darwin, C. 1859. On the origin of the species http://www.talkorigins.org/faqs/origin.html Chapter 3.
3. Wallace, A.R. 1905. *My Life*. New York: Dodd, Mead & Company.
4. Mendel's Paper at MendelWeb. http://www.mendelweb.org/MWpaptoc.html
5. Marchant, J. 1916. *Letters and Reminiscences*, vol **2**. London: Cassell. http://darwin-online.org.uk/converted/published/1916_Marchant_F1592.2.html
6. Dobzhansky, T. 1970. *Genetics of the Evolutionary Process*. New York: Columbia University Press.
7. Fisher, R. 1930. In *The Genetical Theory of Natural Selection*. Oxford at the Clarendon Press. pp 38–39.
8. Pennock, R.T. 2005. "On teaching evolution and the nature of science." In *Evolutionary Science and Society: Educating a New Generation*. J. Cracraft & R. Bybee, Eds.: 7–12. Colorado Springs, CO: BSCS.
9. McClintock, B. 1950. The origin and behavior of mutable loci in maize. *Proc. Natl. Acad. Sci. USA* **36:** 344–355.
10. Lewis, E.B. 1945. The relation of repeats to position effect in drosophila melanogaster. *Genetics* **30:** 137–166.
11. Avery, O.T., C.M. Macleod & M. McCarty. 1944. Studies on the chemical nature of the substance inducing transformation of pneumococcal types: induction of transformation by a desoxyribonucleic acid fraction isolated from pneumococcus type III. *J. Exp. Med.* **79:** 137–158.
12. Benzer, S. 1961. On the topography of the genetic fine structure. *Proc. Natl. Acad. Sci. USA* **47:** 403–415.
13. Mouse Genome Sequencing Consortium. 2002. Initial sequencing and comparative analysis of the mouse genome. *Nature* **420:** 520–562.
14. Caporale, L.H. 2012. Overview of the creative genome: effects of genome structure and sequence on the generation of variation and evolution. *Ann. N. Y. Acad. Sci.* **1267:** 1–10.
15. Crothers, D.M. 2006. "Sequence-dependent properties of DNA and their role in function." In *The Implicit Genome*. L. Caporale, Ed.: 23. Oxford: Oxford University Press
16. Dickerson, R.E. 1983. Base sequence and helix structure variation in B and A DNA. *J. Mol. Biol.* **166:** 419–441.
17. Maizels, N. & L.T. Gray. 2013. The G4 Genome. *PLoS Genet.* **9:** e1003468.
18. Srinivasan, A.R., R.R. Sauers, M.O. Fenley, *et al.* 2009. Properties of the nucleic-acid bases in free and Watson-Crick hydrogen-bonded states: computational insights into the sequence-dependent features of double-helical DNA. *Biophys. Rev.* **1:** 13–20.
19. Sinden, R., V.N. Potaman, E.A. Oussatcheva, *et al.* 2002. Triplet repeat DNA structures and human genetic disease: dynamic mutations from dynamic DNA. *J. Biosci.* **27**(Suppl. 1): 53–65.
20. King, D.G. 2012. Indirect selection of implicit mutation protocols. *Ann. N. Y. Acad. Sci.* **1267:** 45–52.
21. Dickinson, W.J. & J. Seger. 1999. Cause and effect in evolution. *Nature* **399:** 30.
22. Olivera, B.M., M. Watkins, P. Bandyopadhyay, *et al.* 2012. Adaptive radiation of venomous marine snail lineages and the accelerated evolution of venom peptide genes. *Ann. N. Y. Acad. Sci.* **1267:** 61–70.
23. Weiner, J. 1994. *The Beak of the Finch : A Story of Evolution in Our Time*. New York: Knopf.
24. Caporale, L.H. 2006. An Overview of the implicit genome in *The Implicit Genome*. L. Caporale, Ed. Oxford: Oxford University Press.
25. Wilkins, A.S. 2005. Book review. Darwin in the genome: molecular strategies in biological evolution. *Bioessays*, **27:** 111–112.
26. Caporale, L. 2003. Darwin in the genome. *Bioessays* **27:** 984.
27. Caporale, L.H. 2002. *Darwin in the Genome*. New York: McGraw Hill.
28. King, D.G., E.N. Trifonov & Y. Kashi. Tuning Knobs in the Genome: Evolution of Simple Sequence Repeats by Indirect Selection in *The Implicit Genome*. L. Caporale, Ed. Oxford: Oxford University Press.
29. Moxon, R., C. Bayliss & D. Hood. 2006. Bacterial contingency loci: the role of simple sequence dna repeats in bacterial adaptation. *Annu. Rev. Genet.* **40:** 307–333.
30. Gemayel, R., M.D. Vinces, M. Legendre & K.J. Verstrepen. 2010. Variable tandem repeats accelerate evolution of coding and regulatory sequences. *Annu. Rev. Genet.* **44:** 445–477.
31. Van der Ende, A., C.T. Hopman, S. Zaat, *et al.* 1995. Variable expression of class 1 outer membrane protein in Neisseria meningitides is caused by variation in the spacing between the -10 and -35 regions of the promoter. *J. Bacteriol.* **177:** 2475–2480
32. Vinces, M.D., M. Legendre, M. Caldara, *et al.* 2009. Unstable tandem repeats in promoters confer transcriptional evolvability. *Science* **324:** 1213–1216.
33. Fondon, J.W. 3rd & H.R. Garner. 2004. Molecular origins of rapid and continuous morphological evolution. *Proc. Natl. Acad. Sci. USA* **101:** 18058–18063.
34. Fondon, J.W. 3rd, A. Martin, S. Richards, *et al.* 2012. Analysis of microsatellite variation in *Drosophila melanogaster* with population-scale genome sequencing. *PLoS One* **7:** e33036.
35. McIver, L.J., J.F. McCormick, A. Martin, *et al.* 2013. Population-scale analysis of human microsatellites reveals novel sources of exonic variation. *Gene* **516:** 328–334.
36. Caporale, L.H. 2003. Natural selection and the emergence of a mutation phenotype. *Annu. Rev. Microbiol.* **57:** 467–485.
37. Andam, C.P., G.P. Fournier & J.P. Gogarten. 2011. Multilevel populations and the evolution of antibiotic resistance through horizontal gene transfer. *FEMS Microbiol. Rev.* **35:** 756–767.
38. Rodríguez-Trelles, F., R. Tarrío & F.J. Ayala. 2001. Erratic overdispersion of three molecular clocks: GPDH, SOD, and XDH. *Proc. Natl. Acad. Sci. USA* **98:** 11405–11410.
39. Koren, A., P. Polak, J. Nemesh, *et al.* 2012. Differential relationship of DNA replication timing to different forms of

human mutation and variation. *Am. J. Hum. Genet.* **91:** 1033–1040.

40. Walker, S.I., *et al.* 2013. Evolutionary dynamics and information hierarchies in biological systems. *Ann. N.Y. Acad. Sci.* **1305:** 1–17.

41. Durfee, T., A.-M. Hansen, H. Zhi, *et al.* 2008. Transcription profiling of the stringent response in *Escherichia coli. J. Bacteriol.* **190:** 1084–1096.

42. Ciccia, A. & S.J. Elledge. 2010. The DNA damage response: making it safe to play with knives. *Mol. Cell.* **40:** 179–204.

43. Singh, A.H., D.M. Wolf, P. Wang & A.P. Arkin. 2008. Modularity of stress response evolution. *Proc. Natl. Acad. Sci. USA* **105:** 7500–7505.

44. Jones, F.C., M.G. Grabherr, Y.F. Chan, *et al.* 2012. The genomic basis of adaptive evolution in threespine sticklebacks. *Nature* **484:** 55–61.

45. Gally, D.L., J.A. Bogan, B.I. Eisenstein & I.C. Blomfeld. 1993. Environmental regulation of the fim switch controlling type 1 fimbrial phase variation in *Escherichia coli* K-12: effects of temperature and media. *J. Bacteriol.* **175:** 6186–6193.

46. Hsu, E., N. Pulham, L.L. Rumfelt & M.F. Flajnik. 2006. The plasticity of immunoglobulin gene systems in evolution. *Immunol. Rev.* **210:** 8–26.

47. Di Noia, J.M. & M.S. Neuberger. 2007. Molecular mechanisms of antibody somatic hypermutation. *Annu. Rev. Biochem.* **76:** 1–22.

48. Kenter, A.L., S. Feldman, R. Wuerffel, *et al.* 2012. Three-dimensional architecture of the IgH locus facilitates class switch recombination. *Ann. N. Y. Acad. Sci.* **1267:** 86–94.

49. Lucht, J.M., B. Mauch-Mani & H.-Y. Steiner, *et al.* 2002. Pathogen stress increases somatic recombination frequency in *Arabidopsis. Nat. Genet.* **30:** 311–314.

50. McClintock, B. 1984. The significance of responses of the genome to challenge. *Science* **226:** 792–801.

51. Shapiro, J.A. 2010. Mobile DNA and evolution in the 21st century. *Mob. DNA* **1:** 4.

52. Jorgensen, R.A. 2004. Restructuring the genome in response to adaptive challenge: McClintock's bold conjecture revisited. *Cold Spring Harb. Symp. Quant. Biol.* **69:** 349–354.

53. McClintock, B. 1978. "Mechanisms that rapidly reorganize the genome." Proceedings Stadler Genetic Symposium 10: 25-47. In *The Dynamic Genome: Barbara McClintock's Ideas in the Century of Genetics.* N. Fedoroff & D. Botstein, Eds. Cold Spring Harbor Laboratory Press, Plainview, NY 1992.

54. Naito, K., F. Zhang, T. Tsukiyama, *et al.* 2009. Unexpected consequences of a sudden and massive transposon amplification on rice gene expression. *Nature* **461:** 1130–1134.

55. Witkin, E.M. 1967. The radiation sensitivity of *Escherichia coli* B: a hypothesis relating filament formation and prophage induction. *Proc. Natl. Acad. Sci. USA* **57:** 1275–1279.

56. Radman, M. 1975. SOS repair hypothesis: phenomenology of an inducible DNA repair which is accompanied by mutagenesis. *Basic Life Sci.* **5A:** 355–367.

57. Foster, P.L. 2005. Stress responses and genetic variation in bacteria. *Mutat. Res.* **569:** 3–11.

58. Ponder, R.G., N.C. Fonville & S.M. Rosenberg. 2005. A switch from high-fidelity to error-prone DNA double-strand break repair underlies stress-induced mutation. *Mol. Cell.* **16:** 791–804.

59. Shee, C., J.L. Gibson, M.C. Darrow, *et al.* 2011. Impact of a stress-inducible switch to mutagenic repair of DNA breaks on mutation in *Escherichia coli. Proc. Natl. Acad. Sci. USA* **108:** 13659–13664.

60. Al Mamun, A.A., M.J. Lombardo, C. Shee, *et al.* 2012. Identity and function of a large gene network underlying mutagenic repair of DNA breaks. *Science* **338:** 1344–1348.

61. Woese, C.R. & N. Goldenfeld. 2006. How the microbial world saved evolution from the scylla of molecular biology and the charybdis of the modern synthesis. *Proc. Natl. Acad. Sci. USA* **103:** 10696–10701.

62. Alderson, D.L. & J.C. Doyle. 2010. Contrasting views of complexity and their implications for network-centric infrastructures. *IEEE Trans Syst. Man Cybernetics-Part A* **40:** 839–852.

63. Brenner, S. Collinearity and the genetic code 1966. *Proc. Roy. Soc. B.* **164:** 170–180.

64. Caporale, L.H. 1984. Is there a higher level genetic code that directs evolution? *Mol. Cell. Biochem.* **64:** 5–13.

65. Doyle, J., M. Csete & L.H. Caporale. 2006. An engineering perspective: the implicit protocols in *The Implicit Genome.* L. Caporale, Ed. Oxford: Oxford University Press.

66. Doyle, J.C. & M.E. Csete. 2011. Architecture, constraints, and behavior. *Proc. Natl. Acad. Sci. USA* **108**(Suppl. 3): 15624–15630.

67. Vetsigian, K., C. Woese & N. Goldenfeld. 2009. Collective evolution and the genetic code. *Microbiol. Mol. Biol. Rev.* **73:** 14–21.

68. Barry, J.D., J.P. Hall & L. Plenderleith. 2012. Genome hyper-evolution and the success of a parasite. *Ann. N. Y. Acad. Sci.* **1267:** 11–17.

69. Breaker, R.R. 2011. Prospects for Riboswitch discovery and analysis. *Mol. Cell* **43:** 867–879.

70. Prescott, D.M. 1999. The evolutionary scrambling and developmental unscrambling of germline genes in hypotrichous ciliates. *Nucleic Acids Res.* **27:** 1243–1250.

71. Swart, E.C., J.R. Bracht, V. Magrini, *et al.* 2013. The *Oxytricha trifallax* macronuclear genome: a complex eukaryotic genome with 16,000 tiny chromosomes. *PLoS Biol.* **11:** e1001473.

72. Nowacki, M., K. Shetty & L.F. Landweber. 2011. RNA-mediated epigenetic programming of genome rearrangements. *Annu. Rev. Genomics Hum. Genet.* **12:** 367–389.

73. Caporale, L.H. 2000. Mutation is modulated: implications for evolution. *Bioessays* **22:** 388–395.

74. Lambowitz, A.M., N. Craig, R. Gragie & M. Gellert. 2002. *Mobile DNA American Society for Microbiol;* Washington, DC 2nd edition.

75. Lukjancenko, O., T.M. Wassenaar & D.W. Ussery. 2010. Comparison of 61 Sequenced. *Escherichia coli Genomes. Microb. Ecol.* **60:** 708–720.

76. Chandra, F., G. Buzi & J.C. Doyle. 2011. Glycolytic oscillations and limits on robust efficiency. *Science* **333:** 187–192.

77. Yi, T.M., Y. Huang, M.I. Simon & J. Doyle. 2000. Robust perfect adaptation in bacterial chemotaxis through integral feedback control. *Proc. Natl. Acad. Sci. USA* **97:** 4649–4653.

78. Bagheri, N., M.J. Lawson, J. Stelling & F.J. Doyle 3rd. 2008. Modeling the *Drosophila melanogaster* circadian oscillator via phase optimization. *J. Biol. Rhythms* **23:** 525–537.

79. Adler, J., G.L. Hazelbauer & M.M. Dahl. 1973. Chemotaxis toward sugars in *Escherichia coli. J. Bacteriol.* **115:** 824–847.

80. Tsunozaki, M., H.S. Chalasani & C.I. Bargmann. 2008. A behavioral switch: cGMP and PKC signaling in olfactory neurons reverses odor preference in *C. elegans. Neuron* **59:** 959–971.

81. Cisek, P. & J.F. Kalaska. 2010. Neural mechanisms for interacting with a world full of action choices. *Ann. Rev. Neurosci.* **33:** 269–298.

82. Landsteiner, K. 1936. Artificial Conjugated Antigens. Serological Reactions with Simple Chemical Compounds. Chapter V in *The Specificity of Serological Reactions.* Charles C. Thomas: Springfield, Illinois.

83. Turing, A.M. 1942. Report on Cryptographic Machinery Available at Navy Department Washington. http://www.turing.org.uk/sources/washington.html

84. Bertsch McGrayne, S. 2011. *The Theory That Would Not Die How Bayes' Rule Cracked the Enigma Code, Hunted Down Russian Submarines, and Emerged Triumphant from Two Centuries of Controversy.* Yale University Press. New Haven and London.

85. Gjini, E., D.T. Haydon, J.D. Barry & C.A. Cobbold. 2012. The impact of mutation and gene conversion on the local diversification of antigen genes in African trypanosomes. *Mol. Biol. Evol.* **29:** 3321–331.

Ann. N.Y. Acad. Sci. ISSN 0077-8923

Establishing epigenetic domains via chromatin-bound histone modifiers

Fabian Erdel, Katharina Müller-Ott, and Karsten Rippe

Deutsches Krebsforschungszentrum (DKFZ) and BioQuant, Research Group Genome Organization & Function, Im Neuenheimer Feld 280, Heidelberg, Germany

Address for correspondence: Karsten Rippe, Deutsches Krebsforschungszentrum, Research Group Genome Organization & Function (B066), Im Neuenheimer Feld 280, 69120 Heidelberg, Germany. Karsten.Rippe@dkfz.de

The eukaryotic nucleus harbors the DNA genome, which associates with histones and other chromosomal proteins into a complex referred to as chromatin. It provides an additional layer of so-called epigenetic information via histone modifications and DNA methylation on top of the DNA sequence that determines the cell's active gene expression program. The nucleus is devoid of internal organelles separated by membranes. Thus, free diffusive transport of proteins and RNA can occur throughout the space accessible for a given macromolecule. At the same time, chromatin is partitioned into different specialized structures such as nucleoli, chromosome territories, and heterochromatin domains that serve distinct functions. Here, we address the question of how the activity of chromatin-modifying enzymes is confined to chromatin subcompartments. We discuss mechanisms for establishing activity gradients of diffusive chromatin-modifying enzymes that could give rise to distinct chromatin domains within the cell nucleus. Interestingly, such gradients might directly result from immobilization of the enzymes on the flexible chromatin chain. Thus, locus-specific tethering of these enzymes to chromatin could have the potential to establish, maintain, or modulate epigenetic patterns of characteristic domain size.

Keywords: histone modifications; chromatin looping; pattern formation; epigenetics; nuclear organization

Introduction

The genomic DNA of eukaryotes is organized into linear chromosomes with several tens or hundreds of million base pairs (bp) of DNA that is packaged by interactions with histones and other proteins into chromatin. The building block of chromatin is the nucleosome, a complex of a histone octamer that associates with 145–147 bp of DNA wrapped in almost two turns around the octamer. The genome is confined by the nuclear envelope to the cell nucleus. Within the nucleus, no membrane-separated organelles exist. Thus, molecules are free to diffuse within the accessible space. Since diffusion equilibrates concentration gradients, one would expect that all places connected by diffusive transport are equivalent with respect to their molecular composition and function. This is clearly not true for the organization of the genome within the nucleus. Rather, chromatin is organized into functionally

and structurally distinct nuclear subcompartments such as nucleoli, chromosome territories, regions of denser heterochromatin, or more open and frequently more active euchromatin. The associated chromatin states differ with respect to their protein content, nucleosome spacing and positioning, DNA methylation, and histone modifications, as well as the presence of chromatin-associated RNAs. In this manner, access to the DNA for the selection of the active gene-expression program and other genome functions such as DNA replication and repair is controlled (Fig. 1A).

In the presence of freely diffusive enzymes at constant concentrations throughout the nucleus, every nucleosome would have essentially the same probability of colliding productively with an enzyme that could modify it. Modification reactions are characterized by the addition or removal of small chemical groups, such as methyl, acetyl, or phosphate groups, at one of the histone tails or the

doi: 10.1111/nyas.12262

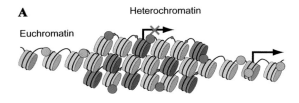

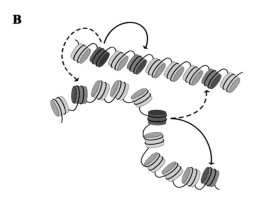

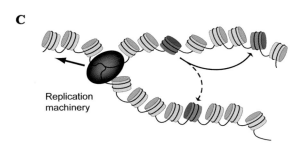

Figure 1. Chromatin states and their propagation. (A) Chromatin is organized into different chromatin states that in a simplified classification are referred to as denser and transcriptionally silenced heterochromatin versus the more open and biologically active euchromatin. The states are different with respect to specific DNA and histone modifications as well as protein composition. In the scheme, DNA methylation is depicted by red circles, nucleosomes with repressive histone modifications such as H3K9me3 or H3K27me3 are shown in green, and nucleosomes with activating modifications (e.g., H3K4me3, H3K36me3) are colored in orange. In addition, bound chromosomal proteins that are associated with the two chromatin states are represented by orange and green circles. (B) Histone modifications can be propagated from nucleation sites to form epigenetic chromatin domains. (C) During DNA replication, nucleosome modifications are lost as new histones are incorporated. This dilution of histone modification patterns behind the replication fork has to be compensated for by propagating the original modification state to the newly formed nucleosomes.

histone core. However, there are several ways to establish heterogeneous chromatin-state patterns in the nucleus: (1) diffusion might be slow as compared to processes that actively create gradients, (2) external boundaries (e.g., the nuclear envelope) can impose mobility constraints that depend on the distance to a given boundary, and (3) a heterogeneous distribution of enzymes that is not balanced by diffusion can result from the binding of enzymes to a less-mobile chromatin scaffold.

Here, we consider different mechanisms for the formation of chromatin subcompartments within the cell nucleus. Although most enzymes that establish such patterns are small enough to move through the whole nucleus, their distribution is not necessarily homogeneous since they can bind to a heterogeneous chromatin network. A direct consequence might be the generation of activity gradients that follow their net abundance. These can lead to the formation of chromatin patterns and thus, partition chromatin into distinct domains.

Epigenetic chromatin states

Historically, chromatin has been globally classified on the basis of the chromatin density distribution in microscopy images into more compact biologically inactive heterochromatin and transcriptionally active euchromatin.[1,2] Transitions between these two states at some chromosomes were accounted for by the introduction of the term *facultative heterochromatin*. One example of this transition is the inactivation of one X chromosome in female mammalian cells, in which one X chromosome adopts a distinct silenced conformation state termed as the Barr body while genes on the other X chromosome are transcriptionally active. Other functionally distinct regions include chromatin at the centromeres,[3,4] pericentromeric heterochromatin,[5,6] chromatin at the nuclear lamina, telomeric chromatin,[7,8] and active and repressive ribosomal genes in nucleolar chromatin.[9]

More recently, systematic chromatin maps have been acquired that evaluate either the protein composition, DNA methylation, or posttranslationally incorporated histone modification patterns such as acetylation, methylation, or phosphorylation to identify structurally and functionally distinct chromatin states.[10–12] These include (1) the identification of five major chromatin states greater than 100 kb in length in *Drosophila*,[10] (2) the categorization of 18 different histone acetylation or methylation marks into nine patterns to characterize functional genomic elements in *Drosophila*,[11] and (3) the evaluation of two histone acetylation marks, six histone methylation modifications, and binding of CCCTC binding factor (CTCF) in different human cell types to identify chromatin patterns that characterize their cell type–specific gene expression profiles.[12]

The functional consequences of establishing a certain chromatin state can be related to changes in DNA accessibility for interacting factors. These can be brought about by different mechanisms. For some modifications such as the acetylation of histones at certain positions (e.g., H4K16ac, H3K56ac, H3K64ac, H3K122ac), there appears to be a direct effect on nucleosome–nucleosome interactions and stability.[13–18] Other modifications enhance the binding of architectural chromatin components that can recognize certain modifications such as methyl-CpG–binding protein MeCP2 for DNA methylation[19] or heterochromatin protein 1 (HP1) for the trimethylation of histone H3 at lysine 9 (H3K9me3) to change chromatin organization.[20] An additional important parameter of the local chromatin structure is the positioning of nucleosomes. These separate the nucleosomal DNA that interacts with histone proteins from the linker DNA between nucleosomes, which is more accessible to soluble factors. This accessibility pattern is tightly linked to histone and DNA modifications as well as other chromatin features.[21] It is functionally important since DNA-dependent processes such as transcription require the binding of enzymes to the DNA. In many instances, binding of transcription factors to nucleosomal DNA is impeded.[22]

Some of the fundamental questions regarding the setting of the cell's active gene expression program by establishing a specific pattern of chromatin states currently remain unanswered: How are chromatin-modifying enzymes targeted to or excluded from chromatin in a spatially defined manner? Once a given nucleosome modification is established, how can it be propagated on the same or different nucleosomes to establish a specific chromatin domain (Fig. 1B)? How is this state reestablished or maintained during DNA replication (Fig. 1C)? Since most enzymes that catalyze DNA and histone modifications are small, they can diffuse rapidly through the nucleus and could potentially modify every nu-

cleosome with which they collide. Thus, spatially heterogeneous epigenetic patterns are established in the context of a well-mixed nucleus. In the following sections, we discuss different possibilities to generate spatially confined chromatin patterns. In particular, we consider the case that chromatin-bound enzymes can give rise to local activity gradients, which appears to be a simple and robust way to establish and maintain epigenetic patterns.

Interactions between genomic loci owing to chromatin dynamics

The contour length of the DNA of one mammalian chromosome is in the order of tens of centimeters, whereas the diameter of the cell nucleus is only 10–20 μm. Thus, DNA packaged into chromatin is highly compacted in the nucleus. Both DNA and the nucleosome chain can be described as polymer chains.[23,24] Individual chromosomes occupy distinct territories during interphase.[25] They have a large friction coefficient and translocate only slowly over micrometer distances within minutes and hours.[26] In contrast, small chromatin domains translocate significantly on length scales of tens to hundreds of nanometers within milliseconds to seconds due to diffusion.[26] Thus, individual segments of the chromatin chain can jiggle around their equilibrium position more rapidly, that is, make small displacements within a local confinement region. This confined diffusive motion of chromatin loci can be directly visualized using different microscopy techniques.[27–29] Motion of chromatin segments within the dense chromatin network leads to collisions, and nucleosomes located on the segments are able to contact each other. This is both true for nucleosomes on the same chromosome and nucleosomes on different chromosomes, with intrachromosomal collisions making the dominant contribution in genome-wide interaction maps (Fig. 1B).[30]

The inherent properties of the chromatin chain define two intrinsic length scales that are relevant for intra- and interchromosomal interactions. The first length scale reflects the flexibility and the conformation of the nucleosome chain. It determines the local concentration profile for nucleosomes on the same chromosome, as discussed in further detail below. This scale dictates the probability for the chain to fold back on itself and thus for local intrachromosomal interactions to occur. The second scale is the confinement radius that restricts the diffusive motion of a chromatin locus in three-dimensional (3D) space. This scale is relevant for interactions that occur between nucleosomes that are in spatial proximity, independent of their genomic coordinate (i.e., their separation distance along the chromosome) and independent of whether they are located on the same chromosome. Measurements of intrachromosomal collision frequencies suggest that intrachromosomal collisions occur efficiently within distances of around 2–4 kb,[31–33] which corresponds to a contour length of several hundred nanometers of the nucleosome chain (Fig. 2). Microscopy experiments with labeled nucleosomes or chromatin loci reveal that the diffusive confinement radius is on a similar length scale of 100–300 nm.[27–29] This fits well with the size of ∼1 Mb topological domains containing ∼5000 nucleosomes that have been identified using the chromosome conformation capture (3C) method[30,34] as well as microscopy-based techniques.[26] Thus, the natural domain size of chromatin might be intimately related to the restricted mobility of chromatin loci, and raises the possibility that the formation of epigenetic domains relies on diffusion-driven mechanisms.

Calculating the contact probability between nucleosomes within the chromatin chain

To make quantitative statements about nucleosome–nucleosome interaction probabilities, polymer models can be applied that are based on either a freely jointed chain[35–37] or a worm-like chain.[38–40] With these theoretical descriptions, the stiffness of a nucleosome chain is described by the statistical segment length or Kuhn length (l) or the persistence length (a), which is related to the Kuhn length according to $l = 2a$. The numerical value of l increases with the stiffness of the polymer. The interaction probability between two sites on the same chain is expressed as the molar local concentration (j_M) of one locus in the proximity of the other.[23,24] The value of j_M is equivalent to the concentration that would be required free in solution to obtain the same contact probability. If a given site is bound by a protein, the same applies for the respective protein concentration. In this case, the occupancy (θ) of the protein-binding site, as well as the interactions with the proteins not bound to the DNA, needs to be considered. In the

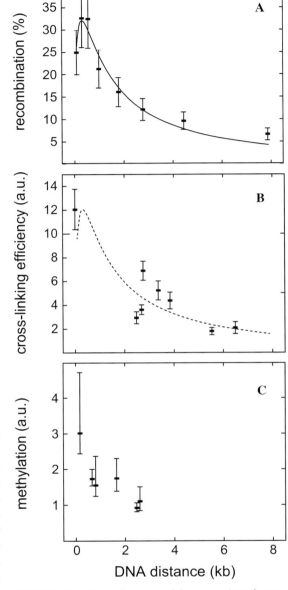

Figure 2. Experimentally measured short-range intrachromosomal interactions. (A) Contact probabilities between two sites on the same chromosome in a human cell line measured by FLP recombination frequency in Ref. 32. The solid line is the least-squared fit to Eq. (3) including some baseline offset and scaling factor. The fit yields $n_b = 160$ bp and $d = 0.16$. (B) Cross-linking efficiency from a 3C experiment.[33] (C) Dependence of *dam* methylation level on the distance of an immobilized *dam* methyltransferase.[31]

balance of this presentation, we assume that the occupancy of the binding site is unity, that is, the interacting site is always fully occupied and c_{free}, the concentration of protein not bound to the DNA, is much smaller than j_{M}. Under these conditions c_{eff}, the effective protein concentration, is equal to j_{M}. If this is not the case, c_{eff} can be calculated according to Eq. 1

$$c_{\text{eff}} = \theta \times j_{\text{M}} + c_{\text{free}}. \tag{1}$$

As described previously,[23,24] j_{M} between two sites at a distance of n segments of length l on a linear polymer can be calculated according to

$$j_{\text{M}}(n) = 0.53 \times n^{-\frac{3}{2}} \times \exp\left(\frac{d-2}{n^2+d}\right)$$
$$\times l^{-3}\frac{\text{mol} \cdot \text{nm}^3}{\text{liter}}. \tag{2}$$

The additional parameter d incorporated into Eq. 2 reduces the contribution of the exponential term if d is larger than zero. This can account for an increase of j_{M} at short separation distances ($n < 4$), for example, due to the size of interacting protein complexes or if intrinsic DNA curvature is present.[24,41] In this general form, the local molar concentration j_{M} of one site in the proximity of the other site is given per l^3, that is, with units mol·nm³/liter. In Eq. 2, the separation distance n between two sites is given as a dimensionless reduced separation distance, which is simply the number of Kuhn segments with length l that corresponds to the site-separation distance. Thus, the expression in Eq. 2 is independent of the characteristics of a specific polymer and has a maximum of $n = 1.6$ for $d = 0$.

For DNA-mediated interactions, it is convenient to express the site-separation distance n by the number of DNA base pairs b. With n_{b} being the number of DNA base pairs per segment length, this yields Eq. 3:

$$j_{\text{M}}(b) = 0.53 \times \left(\frac{b}{n_{\text{b}}}\right)^{-\frac{3}{2}} \times \exp\left(\frac{d-2}{\left(\frac{b}{n_{\text{b}}}\right)^2+d}\right)$$
$$\times l^{-3}\frac{\text{mol} \cdot \text{nm}^3}{\text{liter}}. \tag{3}$$

In order to compute j_{M} from Eq. 3, we need to know the stiffness and contour length of the chain

to be able to express n_{b}. So far no consensus has been reached from experimental measurements for either the stiffness of the nucleosome chain or for its contour length. If expressed in nanometers, both parameters would be highly dependent on the folding of the chain.[42,43] It appears that the short length regime of <10 kb has to be treated separately from long-range interactions at sites 10–500 kb apart. At short separation distances, the interaction probability is likely to reflect the nucleosomal organization of the chain more resembling free double-stranded DNA with intrinsically curved regions due to wrapping of DNA around the histone octamer, and possibly further compaction due to interactions between nucleosomes. Thus, the characteristic contour length of the nucleosome chain per base pair of DNA is very different from that of free DNA, for which a value of 0.34 nm/bp is well accepted.

For the long-range interactions, the higher order organization of chromatin into transient or stable loops of different sizes needs to be included into the polymer model through different stiffness and contour-length parameters.[24,26] Here we focus on the short-range interactions between nucleosomes at distances <10 kb. In general, these appear to be orders of magnitude more frequent than interactions on the 100 kb scale and above, as apparent from 3C experiments that measure both short- and long-range interactions.[30] The experimentally determined short-range intrachromosomal interaction probabilities from three different types of experiments are plotted in Figure 2. Notably, the contact frequencies between two sites on the same chromosome measured by (1) the DNA recombination frequency mediated by the flippase (FLP) enzyme[32] (Fig. 2A), (2) the cross-linking efficiency from 3C experiments[33] (Fig. 2B), and (3) the levels of ectopic adenine methylation around an immobilized *dam* methyltransferase[31] (Fig. 2C) yield very similar distance dependencies for the interaction between two separated sites on the same chromosome. Interactions occur most frequently within <1 kb separation distance and decay to base line levels above ∼5 kb. These conclusions are further supported by *in vitro* experiments with reconstituted nucleosomal arrays and theoretical studies that showed highly efficient enhancer–promoter interactions at separation distances between 0.7 and 4.5 kb.[44]

As the data for FLP recombination frequency were acquired in living mammalian cells at high resolution, they provide an excellent reference for computing j_M independent of the base pair–separation distance. A similar approach was already used in Ref. 32, but with a chain contour length for free DNA. Here, we estimate the contour of an unfolded nucleosome chain from simple geometric considerations as follows. The nucleosome repeat length (i.e., the length of DNA in a nucleosome plus linker DNA) equals roughly 190 bp (e.g., Ref. 22 measured 191 bp for mouse cells). The well-established value of 145–147 bp of nucleosomal DNA results in a linker of 45 bp or 15.3 nm DNA between two nucleosomes. Since the distance between the entry–exit site in the nucleosome is about 9 nm,[45] a contour length of 24.3 nm/191 bp or 0.13 nm/bp is obtained. With these values, a fit of the FLP recombination data yields Eq. 4 with $n_b = 160$ bp, $d = 0.16$, and $l = 160$ bp·0.13 nm/bp $= 21$ nm, for computing the local concentration j_M in mol/liter of a nucleosome in the close proximity of another nucleosome separated by b base pairs on the same chain.

$$j_M(b) = 5.7 \times 10^{-5} \times \left(\frac{b}{160} \right)^{-\frac{3}{2}}$$

$$\times \exp\left(\frac{-1.84}{\left(\frac{b}{160}\right)^2 + 0.16} \right) \frac{\text{mol}}{\text{liter}}. \quad (4)$$

The above description considers the equilibrium distribution for interactions between nucleosomes on the same chain as represented by their local concentrations. The kinetics with which these interactions occur can be estimated from studies of DNA contacts for separation distances of ~10 nm, which corresponds to the length of the linker between nucleosomes. From Brownian dynamics simulations of DNA molecules, collision frequencies of roughly 1000/s were derived.[41] This value fits well with the experimentally determined rates for loop closure of similar-sized DNA hairpins.[46] Furthermore, even much larger chromatin domains show translocations on the 10 and 100 nm length scales in living cells within the 30–50 ms time resolution of the measurements.[27,28] Thus, looping-mediated interactions between nucleosomes occur on the time scale of milliseconds.

Normalization of experimentally measured interaction probabilities

While Eq. 3 is generally applicable for computing interaction probabilities, specific stiffness and contour length parameters of the chain are required to derive the local concentration in molar units. Since there is an ongoing dispute in the field on how to best choose these parameters for chromatin, we introduce a complementary approach to estimate the scaling of the *j*-function for chromatin looping. We consider a single chromosome residing in its territorial space during interphase of the cell cycle. Accordingly, the vast majority of interactions between nucleosomes are intramolecular. This description can be easily extended to the complete genome by considering the nucleus as being filled by a set of chromosomes that occupy distinct spatial territories.[25,26,47]

The local concentration function $(j(n))$ is proportional to the probability $(p(n))$ that a given nucleosome resides within a volume element (dV) around the same chromosome that is separated by distance n along the chain (Fig. 3A). The volume element dV can be described as a cylinder with height dn and radius r, which is chosen as sufficiently small to ensure that the local nucleosome concentration within the volume element is approximately constant. Thus, the average local concentration $\langle j \rangle$ of a given nucleosome within a tube with radius r around the whole chromosome can be expressed as

$$\langle j \rangle = \frac{2}{L} \int_0^{L/2} j(n) \, dn = \frac{1}{L \, \pi \, r^2} = \frac{1}{V_{\text{chr}}}. \quad (5)$$

Here, L is the contour length of the chromosome and $V_{\text{chr}} = L \, \pi \, r^2$ is the volume of the tube surrounding the chromosome, which can be regarded as the volume occupied by the chromosome. For simplicity, the nucleosome at the center of the chromosome was considered for Eq. 5, yielding equal local concentrations to the left and right of the nucleosome. This is a good approximation for all nucleosomes, since mammalian chromosomes have a length of 10–100 Mb while the *j*-functions considered here have a typical width of several thousand base pairs. Using the definition of the average nucleosome concentration $c_{\text{nuc}} = N_{\text{nuc}}/V_{\text{CT}}$, with N_{nuc} being the total number of nucleosomes in the territory with volume V_{CT},

A

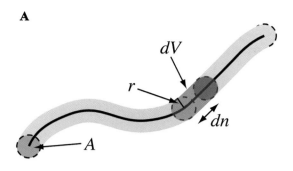

B

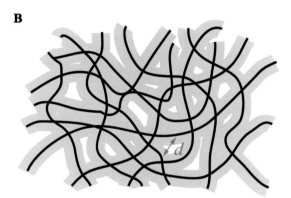

Figure 3. Conversion of interaction probabilities to local concentrations. Experimentally determined contact probabilities can be converted into local concentrations for tightly packed chromatin as found in the cell by normalization to the average nucleosome concentration. (A) The local concentration determines the probability at which a nucleosome may be present in a small volume element dV (red) separated by a given distance along the chromosome. (B) A part of a chromosome territory is depicted. Since chromosomes are tightly packed, the volume of the territory can be similar to the volume of a tube around the chromosome (gray), which has a radius r that is similar to the size of a nucleosome.

Eq. 5 can be rewritten as

$$\langle j \rangle = \frac{1}{V_{chr}} = \frac{V_{CT}}{V_{chr}} \frac{c_{nuc}}{N_{nuc}} \geq \frac{c_{nuc}}{N_{nuc}}. \quad (6)$$

Here, the inequality sign accounts for the fact that a chromosome cannot occupy more space than available in its territory. For a tightly packed chromosome, the two volumes might be similar (i.e., $V_{chr} \approx V_{CT}$). At the average nucleosome concentration $c_{nuc} = 140$ μM measured in a human cell line,[48] the average distance between neighboring nucleosomes corresponds to less than 40 nm if

a random distribution is assumed (Fig. 3B). Thus, the local nucleosome concentration would be approximately constant over the separation distance between neighboring nucleosomes, and the radius r of the chain could be chosen accordingly to ensure that $V_{chr} \approx V_{CT}$. In this case, the average local concentration $\langle j \rangle$ equals the concentration of a nucleosome in the territory $1/V_{CT}$, and Eq. 6 simplifies to

$$N_{nuc} \langle j \rangle \approx c_{nuc}. \quad (7)$$

Thus, Eqs. 6 and 7 provide normalization conditions that are imposed by the constant number of nucleosomes N_{nuc} within the chromosome territory. They can be computed by either integrating the average nucleosome concentration over the nuclear volume or by integrating the local concentration of all nucleosomes over the volume occupied by the chromosome. If one considers j to be proportional to a residence time, Eqs. 6 and 7 mean that a nucleosome spends all its time within the chromosome territory and that its residence times at all the positions it samples add up to this time. Since Eq. 4 is consistent with the normalization according to Eqs. 6 and 7, we conclude that it is justified to use a polymer model with the given parameter set to compute j. With Eqs. 6 and 7 it is possible to estimate absolute concentrations from arbitrary contact probability functions without using a particular polymer model for the description of the nucleosome chain. However, the volume that is occupied by the nucleosome chain has to be estimated carefully. Nonetheless, we feel that the concept described in this section might prove useful for calculating concentrations from experimentally determined interaction maps provided by different methods such as the ones described in the following sections.

Mechanisms for establishing gradients of enzymatic activity

There are several ways to establish gradients in living organisms in the presence of counteracting diffusive mixing. The most straightforward option is a source–sink mechanism, in which a component is rapidly released at one location and rapidly removed at another location. If release and removal are fast compared to the time the components need to diffuse between both locations, a gradient is established. The source–sink model has been discussed extensively in the context of morphogenic gradients during embryogenesis.[49–51] However, for steep intracellular gradients with small spatial extension, this mechanism is inefficient since very high release and removal rates would be needed to counteract diffusion, which is very fast on small length scales. Particles of the size of typical chromatin enzymes require roughly 1 s to diffuse through the whole cell nucleus (i.e., release) and removal processes would have to occur on the millisecond time scale. Although transient intracellular gradients might be established by such a mechanism (e.g., by triggered nuclear import of a protein), it would consume much energy to maintain steep gradients through constitutive pumping.

Another possibility for the establishment of gradients or patterns of enzymatic activity is through binding of the enzyme to a scaffold. In the nucleus, this can be achieved by tethering enzymes to the nuclear membrane or to chromatin (Fig. 4). Since chromatin fibers exhibit confined diffusion (i.e., they jiggle around their equilibrium position but do not make large translocations most of the time), binding of an enzyme to a chromatin locus increases the local enzyme concentration in the vicinity of the locus. Similarly, proteins attached to the nuclear envelope might diffuse laterally but do not make large radial translocations, since the radial position of the membrane is fixed by the nuclear lamina (Fig. 4A). Thus, a steep radial gradient can easily be established by tethering an enzyme to the nuclear envelope. The spatial distribution of an enzyme might directly translate into a distribution of enzymatic activity if the enzyme is active in the bound state. A well-studied example for such a case is the immobilization of regulator of chromosome condensation 1 (RCC1) on mitotic chromosomes[52] (Fig. 4B). RCC1 serves as guanine nucleotide-exchange factor for the small GTPase Ran. RCC1 is even more active in the chromatin-bound state,[53] which leads to enhanced production and release of Ran–GTP (guanosine triphosphate) at mitotic chromosomes. Although Ran–GTP can quickly diffuse away from the chromosomes, which would ultimately result in a uniform Ran–GTP distribution, a constitutive gradient is achieved since Ran–GTP has a half-life that is shorter than the respective diffusion time.[54,55] Consequently, Ran–GTP quickly converts into Ran–GDP (guanosine diphosphate) after having detached from chromatin, leading to increased Ran–GTP levels around

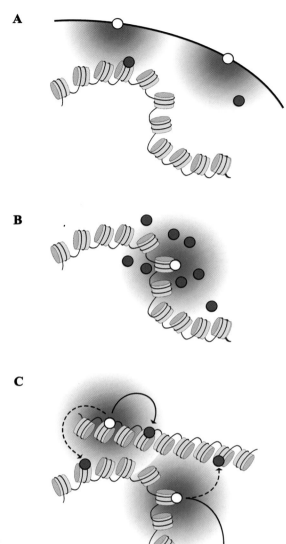

Figure 4. Pattern-formation mechanisms involving immobilized enzymes. Patterns or gradients within the cell nucleus can be established by different mechanisms: (A) Binding of enzymes (white circles) at the nuclear membrane where they continuously catalyze reactions to generate products (blue circles). (B) Binding of enzymes to the nucleosome chain, which produces compounds that diffuse away until they decay. (C) Binding of chromatin-modifying enzymes to the nucleosome chain, which can interact with neighboring nucleosomes to modify them via looping-mediated interactions to establish a specific pattern of epigenetic modifications around the binding site.

chromosomes. In this manner, a persistent Ran–GTP gradient is established around chromatin-bound RCC1.

Other chromatin-bound enzymes could establish similar gradients and patterns. Enzymes that are bound to chromatin and at the same time use chromatin as their substrate might establish distinct chromatin states around their binding sites (Fig. 4C). For patterns of soluble substrates, it is important to provide an efficient depletion mechanism, which would be the short half-life in the case of Ran–GTP. This is not required if chromatin is the substrate, since nucleosomes exhibit only confined diffusion and thus cannot balance concentration gradients of a nucleosomal species on biologically relevant time scales. If an enzyme immobilized on the chromatin fiber is able to productively collide with adjacent substrate nucleosomes as described in the previous section, this would immediately give rise to a higher concentration of enzymatic activity in the vicinity and thus to an enrichment of modified nucleosomes. In this scenario, the size of such a domain around the immobilized enzyme is strongly influenced by the flexibility of the chromatin chain. The folding and compaction state of the fiber could result in a modulation of domain size. With respect to setting DNA or histone modifications by a stably chromatin-bound enzyme that catalyzes the corresponding reaction, several predictions can be made. First, the chromatin modification is established locally around a nucleation region if the free concentration of the catalytically active form of the nucleated enzyme is comparably small with respect to the local concentration due to chromatin looping. Otherwise, the effect would no longer be localized, since the reaction would also be catalyzed at other sites by the freely mobile enzyme. Second, the typical spreading distance is determined by the diffusive motion of the nucleation region as well as the concentration (and productivity) of the free enzymes. For typical values, domain sizes of less than 10 kb are expected. Third, no boundary factors are required to limit the spreading of the modifications, since the diffusive motion of the bound enzyme is inherently confined. Fourth, modified chromatin regions have spherical shapes since there is no preferred direction for diffusive motion. And fifth, the steady-state modification level is not bistable unless nucleation sites are coupled in a complex manner via feedback loops.

Although it is difficult to directly demonstrate that the propagation of epigenetic modifications operates by such a mechanism, it is supported by experimental studies in various model systems, as discussed later.

A number of other mechanisms for chromatin pattern formation in well-mixed systems can be envisioned. One example is Turing patterns that can emerge as self-organizing structures due to different diffusion coefficients of two counteracting enzymes. These have been discussed in the context of biological systems such as pattern formation in animal skin development.[56] Although Turing patterns could in principle play a role in intracellular pattern formation, they are likely to lack the spatial precision to provide a robust mechanism by which the cell would be able to control the formation of epigenetic chromatin signatures. Loci that are to be modified would have to be positioned rather accurately with respect to each other and with respect to diffusive boundaries to avoid misregulation of gene activity.

Another interesting pattern formation mechanism includes Ising-type models based on nearest-neighbor interactions. Classically, the Ising model was used to describe ferromagnets that contain distinct magnetic domains, but it was also applied to model biological systems such as activity patterns in neural networks.[57] Recently, models based on nearest-neighbor interactions have also been used to describe epigenetic patterns.[58,59] In these models, an enzyme preferentially modifies a nucleosome that has a modified neighbor on the same chain, resulting in linear spreading of the modification along the chromosome. Although in simulations these models have been found to produce finite domains around a nucleation site, it seems challenging for the cell to robustly define the position and the size of the domain using a linear-spreading model. In particular, unlimited spreading at the boundary of the domain has to be prevented within the noisy cellular environment, in which protein concentrations and occupancies of binding sites fluctuate. To efficiently realize a nearest-neighbor model on chromatin, the association rate of the modifying enzyme has to be much higher for binding to a modified nucleosome than to a potential substrate. Otherwise, the enzyme could directly modify a substrate nucleosome instead of binding to a modified one and subsequently modifying the neighbor. To our knowledge, such behavior has not been reported experimentally. Moreover, it is elusive how an enzyme would discriminate between a nucleosome on the same chain and a nucleosome on a different chain if both are located at similar spatial distances in the crowded environment of the nucleus. Thus, it will be interesting to see if a molecular basis for such models will be identified in the future.

Finally, more complex hybrid mechanisms for establishing epigenetic patterns on chromatin were proposed that involve the integration of symmetrical positive-feedback loops in which nucleosomes are actively modified by proteins that bind to a given histone mark and, at the same time, can interact with proteins that set or remove histone modifications.[60] In the latter model, the enzymes can both act on neighboring nucleosomes and exert some more long-range interactions with nucleosomes at a distance. To limit the spreading of a distinct modification mark, boundary elements were introduced. A similar model was used by Angel *et al.*[61] This type of theoretical description results in bistable chromatin states, that is, for the locus under consideration two distinct states can stably coexist, which could correspond to either transcriptional activity or transcriptional silencing.[62]

Features of experimentally observed epigenetic patterns

As discussed earlier, confined diffusion of chromatin-bound epigenetic modifiers could give rise to localized finite epigenetic domains. The corresponding modification profiles are expected to follow the intrachromosomal contact probabilities depicted in Figure 2. Thus, a chromosome-bound enzyme would modify nucleosomes on the same chromosome within ~4 kb around its binding site. These expected domain sizes agree very well with experimentally determined histone-methylation profiles, in which the modification was induced by artificially tethering a protein to a specific locus. One example is the recruitment of histone methyltransferase Clr4 to three adjacent *GAL* sites in yeast that results in the histone H3 lysine 9 dimethylation (H3K9me2) profile depicted in Figure 5A.[63] In another study, Hathaway *et al.* induced gene silencing by artificially recruiting HP1 in mouse embryonic stem cells and fibroblasts.[58] They observed spreading of the repressive H3K9me3 modification around the nucleation site with smoothly decreasing bor-

ders (Fig. 5B). In both studies, the histone modification profiles around the locally chromatin-tethered protein were very similar to the local concentration profiles predicted according to Eq. 4 (Fig. 5C). A similarly shaped H3 lysine 27 trimethylation domain was found for the polycomb repressive complex 2 (PRC2)-based silencing of the floral repressor locus C (*FLC*) gene.[61] Thus, the overall shape of histone methylation domains observed in these experiments can be explained as originating from a chromatin-bound enzyme that propagates the modification via chromatin dynamics along the chain. It is noted that the exact size of a domain generated in this manner will also depend on the concentration and activity of the unbound enzyme as well as counteracting enzymes that remove a given histone modification (Fig. 5D–F). In particular, domains will only be formed if the productive collision frequency with immobilized enzymes is significantly higher than modifications catalyzed by the freely diffusive enzymes. In addition to an increase of the local concentration in the vicinity of the chromatin-bound protein, as reflected in the value of j_M, chromatin binding could also involve allosteric activation of the enzyme (e.g., due to multimerization to increase its spatially confined activity). For example, the tethering of the bacterial enhancer–binding protein NtrC to DNA is accompanied by its multimerization to create an active complex that would not form freely in solution and that interacts with RNA polymerase at the promoter through DNA looping.[64,65] Another potential layer of regulation is the accessibility of the enzymes to different chromatin domains. In case of a significant size difference between a modifying enzyme and its antagonist, densely packed chromatin regions might have a bias for one of the counteracting activities, since only the smaller enzyme can easily access such a region. For example, the histone H3K9 methyltransferases Suv39h1, G9a, GLP, and SETDB1 were found to associate into a complex of MDa molecular weight.[66] In contrast, a counteracting H3K9me2 demethylation activity for an enzyme from the Jumonji family was present in a complex of only 300 kDa.[67]

The size of the histone modification domain established from a chromatin-bound enzyme is modulated by the frequency of productive collisions of a nucleosome with the counteracting enzyme (Fig. 5E and F). If productive collisions with free modifiers occur rarely and collisions with the

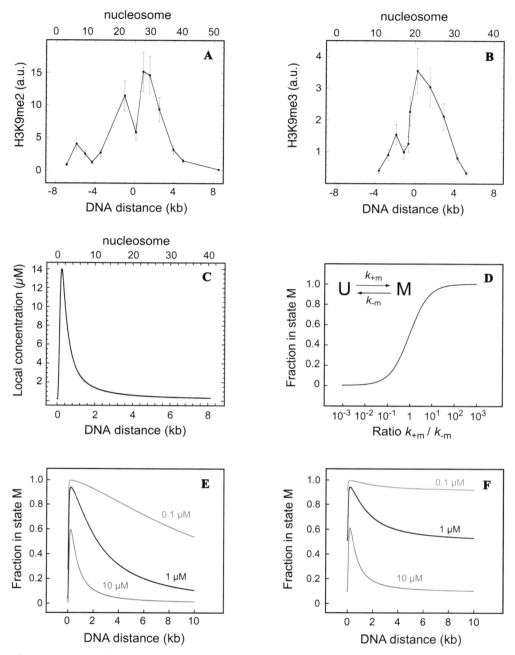

Figure 5. Experimental and theoretical one-dimensional histone modification domain profiles along the nucleosome chain. (A) Artificial recruitment of Clr4, the H3K9-specific methyltransferase in yeast, to three adjacent sites (*GAL* sites) in euchromatin resulted in a symmetrically distributed H3K9 dimethylation profile.[63] (B) Artificial tethering of HP1 to a site in the *Oct-4* promoter in mouse cells induced H3K9 trimethylation in the local vicinity.[58] (C) Calculated local concentrations according to Eq. 4. (D) Histone modifications at a given nucleosome arise from the opposing activities of enzymes that catalyze the addition or removal of the modification M with rates k_{+m} and k_{-m}, respectively, according to the relation $M = 1/(1+k_{-m}/k_{+m}) = k_{+m}/(k_{+m} + k_{-m})$. (E) Histone-modification domains with different spatial extensions can be formed by a mechanism that relies on nucleosomal collisions within the chain. The domain size can be modulated through the concentration and activity of free and bound enzymes. Plots for the indicated concentrations of counteracting soluble enzymes are shown in the absence of free modifying enzyme. (F) Same as in panel E but for a 1 μM concentration of free modifying enzyme.

counteracting enzyme occur more frequently (but less frequently than collisions with the immobilized modifiers), modification profiles that resemble the experimental contact-probability distributions depicted in Figure 5A and B will be obtained. However, for other cases, the relevant parameters might be very different (Fig. 5E and F). If the activity of counteracting processes can be neglected, relatively large domains might be formed. An example for modification domains established in the absence of counteracting processes are the patterns around ectopic *dam* methyltransferase molecules tethered to chromatin or the nuclear lamina. These domains span 2–3 kb on the same chromosome (Fig. 2C) and extend up to 1 μm from the lamina in 3D space.[31,68] Thus, the cell might be able to establish epigenetic domains up to the 1-μm length scale by adjusting the amount and binding affinity of chromatin-modifying enzymes and their antagonists. The underlying diffusion-driven mechanism simply relies on the intrinsic flexibility of chromatin and does not require additional boundaries to restrict spreading of a given modification. The maximum propagation rate according to this mechanism is limited by the collision frequency between nucleosomes, which is around 1000/s, as described earlier. Most epigenetic modifiers seem to have rather low modification rates, with some minutes for histone acetylation (see Ref. 69 and references therein) and up to hours for histone methylation.[70] Consistent with this view, the spreading rate measured in living cells for H3K9 trimethylation is rather slow at ∼0.18 nucleosomes/h.[58] Thus, the collision frequency between nucleosome substrates and chromatin-bound modifiers does probably not represent a rate-limiting step for epigenetic pattern formation.

Epigenetic memory

A long-standing question in epigenetics is how histone modifications are inherited (i.e., transmitted through genome replication and cell division). For DNA methylation, this is accomplished by a dedicated machinery that recognizes hemi-methylated DNA after replication and reestablishes the fully methylated state.[71] For histone modifications, no comparable duplication mechanism is known and the cell has to cope with the enormous combinatorial variety that arises from at least 80 potential modification sites on histone H3 and H4 that are sub-ject to acetylation, mono-, di- and trimethylation, phosphorylation, etc. In the model proposed here, histone modifications could simply be transmitted through replication by chromatin-bound enzymes that collide with nucleosomes on both daughter helices (Fig. 1C). This mechanism is compatible with an arbitrary distribution of newly incorporated nucleosomes behind the replication fork, since it is irrelevant on which chain a given modifying enzyme is immobilized. Each chromatin-bound enzyme could establish a modification domain in its spatial proximity via both intra- and interfiber collisions, yielding two fibers with similar patterns. To ensure that the amount of chromatin-bound enzyme is invariant, the density of nucleation sites has to be kept constant. Such nucleation sites could be made up of a specific DNA sequence and/or DNA methylation, which are both retained during replication. In the presence of positive feedback, the propagated histone modification might itself contribute to establish additional nucleation sites. Furthermore, nascent RNA transcripts that originate from a defined locus and bind chromatin-modifying enzymes are candidates for setting up nucleation sites. We find such a simple collision-driven inheritance mechanism very attractive, but note that further experimental investigations are required to demonstrate its existence and molecular details for a given histone modification.

Conclusions

It is a fascinating question how cells manage to establish numerous subcompartments in the nucleus given that diffusion balances concentration gradients of all soluble factors. Since most enzymes that establish localized chromatin states are small enough to diffuse rapidly through the complete cell nucleus, the question arises why epigenetic modifications of nucleosomes and DNA do not display homogeneous distributions that reflect diffusive collisions with the corresponding enzymes within the well-mixed nucleoplasm. Here, we propose a simple mechanism that can explain how epigenetic domains are formed around an enzyme that is bound to chromatin. Random motions of the enzyme together with the chromatin segment to which it is bound will lead to collisions with nucleosomes in spatial proximity and an elevated modification probability in a confined region around the binding site. As described above, many experimental findings are consistent with such a mechanism for establishing epigenetic patterns. According to our model, the patterns cannot form spontaneously at an arbitrary position but originate from a nucleation site at which the modifying enzyme is immobilized. Nucleation sites might be formed by special sequence elements, modified DNA bases such as 5-methylcytosine, modified histone residues, or chromatin regions with a specific composition of proteins and RNA. Thus, potential patterns would be imprinted in the chromatin polymer as a distribution of these nucleation sites, which can be bound and interpreted if the appropriate adaptor molecules are present. On the one hand, this ensures robustness, since patterns cannot emerge spontaneously at the wrong sites and cannot extend erroneously into regions where they should not be. On the other hand, the activation of imprinted nucleation sites depends on macromolecules that bind them, that is, patterns can be switched by modulating the expression level, the intracellular localization, or the binding behavior of these molecules. Upon some stimulus, the cell can bring the appropriate adaptor molecule into the nucleus, which then leads to the activation of the respective nucleation sites and the formation of a given epigenetic pattern. An example would be a hormone receptor that is imported into the nucleus upon hormone exposure. Thus, a rather static distribution of nucleation sites that is stably imprinted into chromatin of a given cell type can be interpreted in a dynamic fashion to endow the cell with sufficient plasticity.

The formation of extended epigenetic patterns as opposed to the modulation of single genes might be beneficial for the cell to increase the robustness of gene regulation. First, genes that are located in spatial proximity can easily be coregulated, which introduces some kind of modularity into the collection of gene expression programs. If genes were targeted individually, all binding affinities at their regulatory sequences would have to be the same to ensure equal dose–response curves, and stochastic effects that could arise from low numbers of activators binding only to a subset of target genes would become critical. Second, single mutations in binding sites could lead to complete deregulation of a particular gene but would not abolish the formation of a local epigenetic pattern if a redundant subset of nucleation sites was present. This might enhance cellular tolerance with respect to mutations. Third, a

memory effect is achieved if a histone modification with low turnover is established, i.e., gene expression can be regulated on the desired time scale independently of the duration of a given stimulus. This is useful for modulating the strength of responses to transient stimuli.

The local propagation mechanism described here is not limited to linear progression along a chromatin fiber but happens in three dimensions, since there is no preferred direction of the diffusive motion of the nucleosome chain. Accordingly, it provides a straightforward explanation for the approximately spherical shape of macroscopic modification patterns.[72,73] This is different from models that either involve a linear-spreading mechanism along the DNA that exclusively targets residues adjacent to methylated ones which can be modified[58,59] or that implicitly involve a looping of the nucleosome chain to allow for interaction between distant regions but invoke the existence of so-called boundary or insulator elements to limit spreading.[60] The mechanism described here has a simple molecular basis and might be applicable to many cellular processes. Furthermore, from an evolutionary point of view it seems simpler to develop a patterning mechanism that relies on the intrinsic flexibility of chromatin compared to the coevolution of additional mechanisms that are responsible for positioning boundaries. Thus, we anticipate that patterning mechanisms driven by the localized propagation of epigenetic marks via chromatin dynamics and bound modifiers will prove to be relevant for our understanding of chromatin biology.

Acknowledgments

K.R. is grateful to the organizers of the workshop "Evolutionary Dynamics and Information Hierarchies" that was held on August 19–September 9, 2012 at the Aspen Center for Physics in Aspen, Colorado, where part of this work was conducted. Furthermore, we thank Katharina Deeg for critically reading the manuscript and Vladimir Teif and Thomas Höfer for discussion, and acknowledge support within the project *EpiSys* by the German Federal Ministry of Education and Research (BMBF) in the SysTec program.

Conflicts of interest

The authors declare no conflicts of interest.

References

1. Grewal, S.I. & S. Jia. 2007. Heterochromatin revisited. *Nat. Rev. Genet.* **8:** 35–46.

2. Eissenberg, J.C. & G. Reuter. 2009. Cellular mechanism for targeting heterochromatin formation in Drosophila. *Int. Rev. Cell Mol. Biol.* **273:** 1–47.

3. Allshire, R.C. & G.H. Karpen. 2008. Epigenetic regulation of centromeric chromatin: old dogs, new tricks? *Nat. Rev. Genet.* **9:** 923–937.

4. Black, B.E. & D.W. Cleveland. 2011. Epigenetic centromere propagation and the nature of CENP-a nucleosomes. *Cell* **144:** 471–479.

5. Probst, A.V. & G. Almouzni. 2008. Pericentric heterochromatin: dynamic organization during early development in mammals. *Differentiation* **76:** 15–23.

6. Maison, C. *et al.* 2010. Heterochromatin at mouse pericentromeres: a model for de novo heterochromatin formation and duplication during replication. *Cold Spring Harbor Symp. Quant. Biol.* **75:** 155–165.

7. Blasco, M.A. 2007. The epigenetic regulation of mammalian telomeres. *Nat. Rev. Genet.* **8:** 299–309.

8. de Lange, T., V. Lundblad & E. Blackburn. 2006. *Telomeres.* Cold Spring Harbor: Cold Spring Harbor Laboratory Press.

9. Grummt, I. & G. Langst. 2013. Epigenetic control of RNA polymerase I transcription in mammalian cells. *Biochim. Biophys. Acta.* **1829:** 393–404.

10. Filion, G.J. *et al.* 2010. Systematic protein location mapping reveals five principal chromatin types in Drosophila cells. *Cell* **143:** 212–224.

11. Kharchenko, P.V. *et al.* 2011. Comprehensive analysis of the chromatin landscape in Drosophila melanogaster. *Nature* **471:** 480–485.

12. Ernst, J. *et al.* 2011. Mapping and analysis of chromatin state dynamics in nine human cell types. *Nature* **473:** 43–49.

13. Shogren-Knaak, M. *et al.* 2006. Histone H4-K16 acetylation controls chromatin structure and protein interactions. *Science* **311:** 844–847.

14. Robinson, P.J.J. *et al.* 2008. 30 nm chromatin fibre decompaction requires both H4-K16 acetylation and linker histone eviction. *J. Mol. Biol.* **381:** 816–825.

15. Daujat, S. *et al.* 2009. H3K64 trimethylation marks heterochromatin and is dynamically remodeled during developmental reprogramming. *Nat. Struct. Mol. Biol.* **16:** 777–781.

16. Watanabe, S. *et al.* 2013. A histone acetylation switch regulates H2A.Z deposition by the SWR-C remodeling enzyme. *Science* **340:** 195–199.

17. Yang, D. & G. Arya. 2011. Structure and binding of the H4 histone tail and the effects of lysine 16 acetylation. *Phys. Chem. Chem. Phys.* **13:** 2911–2921.

18. Allahverdi, A. *et al.* 2011. The effects of histone H4 tail acetylations on cation-induced chromatin folding and self-association. *Nucleic Acids Res.* **39:** 1680–1691.

19. Adkins, N.L. & P.T. Georgel. 2011. MeCP2: structure and function. *Biochem. Cell Biol.* **89:** 1–11.

20. Thiru, A. *et al.* 2004. Structural basis of HP1/PXVXL motif peptide interactions and HP1 localisation to heterochromatin. *EMBO J.* **23:** 489–499.

21. Erdel, F. *et al.* 2011. Targeting chromatin remodelers: signals and search mechanisms. *Biochim. Biophys. Acta.* **1809:** 497–508.

22. Teif, V.B. *et al.* 2012. Genome-wide nucleosome positioning during embryonic stem cell development. *Nat. Struct. Mol. Biol.* **19:** 1185–1191.

23. Rippe, K., P.H. von Hippel & J. Langowski. 1995. Action at a distance: DNA-looping and initiation of transcription. *Trends Biochem. Sci.* **20:** 500–506.

24. Rippe, K. 2001. Making contacts on a nucleic acid polymer. *Trends Biochem. Sci.* **26:** 733–740.

25. Cremer, T. & M. Cremer. 2010. Chromosome territories. *Cold Spring Harbor Perspect. Biol.* **2:** a003889.

26. Strickfaden, H., T. Cremer & K. Rippe. 2012. "Higher order chromatin organization and dynamics." In *Genome Organization and Function in the Cell Nucleus.* K. Rippe, Ed.: 417–447. Weinheim: Wiley-VCH.

27. Jegou, T. *et al.* 2009. Dynamics of telomeres and promyelocytic leukemia nuclear bodies in a telomerase negative human cell line. *Mol. Biol. Cell.* **20:** 2070–2082.

28. Levi, V. *et al.* 2005. Chromatin dynamics in interphase cells revealed by tracking in a two-photon excitation microscope. *Biophys. J.* **89:** 4275–4285.

29. Hihara, S. *et al.* 2012. Local nucleosome dynamics facilitate chromatin accessibility in living mammalian cells. *Cell Rep.* **2:** 1645–1656.

30. Dekker, J., M.A. Marti-Renom & L.A. Mirny. 2013. Exploring the three-dimensional organization of genomes: interpreting chromatin interaction data. *Nat. Rev. Genet.* **14:** 390–403.

31. van Steensel, B. & S. Henikoff. 2000. Identification of *in vivo* DNA targets of chromatin proteins using tethered dam methyltransferase. *Nat. Biotechnol.* **18:** 424–428.

32. Ringrose, L. *et al.* 1999. Quantitative comparison of DNA looping *in vitro* and *in vivo*: chromatin increases effective DNA flexibility at short distances. *EMBO J.* **18:** 6630–6641.

33. Dostie, J. *et al.* 2006. Chromosome Conformation Capture Carbon Copy (5C): a massively parallel solution for mapping interactions between genomic elements. *Genome Res.* **16:** 1299–1309.

34. Dixon, J.R. *et al.* 2012. Topological domains in mammalian genomes identified by analysis of chromatin interactions. *Nature* **485:** 376–380.

35. Kuhn, W. 1934. Über die Gestalt fadenförmiger Moleküle in Lösungen. *Koll. Z.* **68:** 2–15.

36. Jacobson, H. & W.H. Stockmayer. 1950. Intramolecular reaction in polycondensations. I. The theory of linear systems. *J. Chem. Phys.* **18:** 1600–1606.

37. Flory, P.J. 1969. *Statistical Mechanics of Chain Molecules.* New York: Wiley.

38. Kratky, O. & G. Porod. 1949. Röntgenuntersuchung gelöster Fadenmoleküle. *Rec. Trav. Chim.* **68:** 1106–1113.

39. Shimada, J. & H. Yamakawa. 1984. Ring-closure probabilities of twisted wormlike chains. Application to DNA. *Macromolecules* **17:** 689–698.

40. Becker, N.B., A. Rosa & R. Everaers. 2010. The radial distribution function of worm-like chains. *Eur. Phys. J. E. Soft Matter.* **32:** 53–69.

41. Merlitz, H. *et al.* 1998. Looping dynamics of linear DNA molecules and the effect of DNA curvature: a study by Brownian dynamics simulation. *Biophys. J.* **74:** 773–779.

42. Rippe, K. 2012. "The folding of the nucleosome chain." In *Genome Organization and Function in the Cell Nucleus.* K. Rippe, Ed.: 139–167. Weinheim: Wiley-VCH.

43. Rippe, K., R. Stehr & G. Wedemann. 2012. "Monte Carlo simulations of nucleosome chains to identify factors that control DNA compaction and access." In *Innovations in Biomolecular Modeling and Simulation.* T. Schlick, Ed.: 198–235. Cambridge: RSC Publishing.

44. Kulaeva, O.I. *et al.* 2012. Internucleosomal interactions mediated by histone tails allow distant communication in chromatin. *J. Biol. Chem.* **287:** 20248–20257.

45. Kepper, N. *et al.* 2011. Force spectroscopy of chromatin fibers: extracting energetics and structural information from Monte Carlo simulations. *Biopolymers* **95:** 435–447.

46. Bonnet, G., O. Krichevsky & A. Libchaber. 1998. Kinetics of conformational fluctuations in DNA hairpin-loops. *Proc. Natl. Acad. Sci. USA* **95:** 8602–8606.

47. Cremer, T. & C. Cremer. 2001. Chromosome territories, nuclear architecture and gene regulation in mammalian cells. *Nat. Rev. Genet.* **2:** 292–301.

48. Weidemann, T. *et al.* 2003. Counting nucleosomes in living cells with a combination of fluorescence correlation spectroscopy and confocal imaging. *J. Mol. Biol.* **334:** 229–240.

49. Crick, F. 1970. Diffusion in embryogenesis. *Nature* **225:** 420–422.

50. Yu, S.R. *et al.* 2009. Fgf8 morphogen gradient forms by a source-sink mechanism with freely diffusing molecules. *Nature* **461:** 533–536.

51. Driever, W. & C. Nusslein-Volhard. 1988. A gradient of bicoid protein in Drosophila embryos. *Cell* **54:** 83–93.

52. Carazo-Salas, R.E. *et al.* 1999. Generation of GTP-bound Ran by RCC1 is required for chromatin-induced mitotic spindle formation. *Nature* **400:** 178–181.

53. Nemergut, M.E. *et al.* 2001. Chromatin docking and exchange activity enhancement of RCC1 by histones H2A and H2B. *Science* **292:** 1540–1543.

54. Klebe, C. *et al.* 1995. Interaction of the nuclear GTP-binding protein Ran with its regulatory proteins RCC1 and RanGAP1. *Biochemistry* **34:** 639–647.

55. Hinkle, B. *et al.* 2002. Chromosomal association of Ran during meiotic and mitotic divisions. *J. Cell Sci.* **115:** 4685–4693.

56. Maini, P.K., R.E. Baker & C.M. Chuong. 2006. Developmental biology. The Turing model comes of molecular age. *Science* **314:** 1397–1398.

57. Hopfield, J.J. 1982. Neural networks and physical systems with emergent collective computational abilities. *Proc. Natl. Acad. Sci. USA* **79:** 2554–2558.

58. Hathaway, N.A. *et al.* 2012. Dynamics and memory of heterochromatin in living cells. *Cell* **149:** 1447–1460.

59. Hodges, C. & G.R. Crabtree. 2012. Dynamics of inherently bounded histone modification domains. *Proc. Natl. Acad. Sci. U. S. A.* **109:** 13296–13301.

60. Dodd, I.B. *et al.* 2007. Theoretical analysis of epigenetic cell memory by nucleosome modification. *Cell* **129:** 813–822.

61. Angel, A. *et al.* 2011. A Polycomb-based switch underlying quantitative epigenetic memory. *Nature* **476:** 105–108.

62. Ferrell, J.E. 2012. Bistability, bifurcations, and Waddington's epigenetic landscape. *Curr. Biol.* **22:** R458–R466.

63. Kagansky, A. *et al.* 2009. Synthetic heterochromatin bypasses RNAi and centromeric repeats to establish functional centromeres. *Science* **324:** 1716–1719.

64. Rippe, K., N. Mücke & A. Schulz. 1998. Association states of the transcription activator protein NtrC from *E. coli* determined by analytical ultracentrifugation. *J. Mol. Biol.* **278:** 915–933.

65. Schulz, A., J. Langowski & K. Rippe. 2000. The effect of the DNA conformation on the rate of NtrC activated transcription of *E. coli* RNA Polymerase s^{54} holoenzyme. *J. Mol. Biol.* **300:** 709–725.

66. Fritsch, L. *et al.* 2010. A subset of the histone H3 lysine 9 methyltransferases Suv39h1, G9a, GLP, and SETDB1 participate in a multimeric complex. *Mol. Cell* **37:** 46–56.

67. Yamane, K. *et al.* 2006. JHDM2A, a JmjC-containing H3K9 demethylase, facilitates transcription activation by androgen receptor. *Cell* **125:** 483–495.

68. Kind, J. *et al.* 2013. Single-cell dynamics of genome-nuclear lamina interactions. *Cell* **153:** 178–192.

69. Görisch, S.M. *et al.* 2005. Histone acetylation increases chromatin accessibility. *J. Cell Sci.* **118:** 5825–5834.

70. Zee, B.M. *et al.* 2010. *In vivo* residue-specific histone methylation dynamics. *J. Biol. Chem.* **285:** 3341–3350.

71. Law, J.A. & S.E. Jacobsen. 2010. Establishing, maintaining and modifying DNA methylation patterns in plants and animals. *Nat. Rev. Genet.* **11:** 204–220.

72. Bewersdorf, J., B.T. Bennett & K.L. Knight. 2006. H2AX chromatin structures and their response to DNA damage revealed by 4Pi microscopy. *Proc. Natl. Acad. Sci. USA* **103:** 18137–18142.

73. Eck, S. *et al.* 2013. Segmentation of heterochromatin foci using a 3D spherical harmonics intensity model. In *Bildverarbeitung für die Medizin 2013*, H.-P. Meinzer *et al.*, Eds.: 308–313. Berlin: Springer.

Ann. N.Y. Acad. Sci. ISSN 0077-8923

ANNALS OF THE NEW YORK ACADEMY OF SCIENCES

Issue: *Evolutionary Dynamics and Information Hierarchies in Biological Systems*

The molecular basis for the development of neural maps

Yi Wei, Dmitry Tsigankov, and Alexei Koulakov

Cold Spring Harbor Laboratory, Cold Spring Harbor, New York

Address for correspondence: Alexei Koulakov, CSHL-NA, 1 Bungtown Road, Cold Spring Harbor, New York 11724. akula@cshl.edu

Neural development leads to the establishment of precise connectivity in the nervous system. By contrasting the information capacities of cortical connectivity and the genome, we suggest that simplifying rules are necessary in order to create cortical connections from the limited set of instructions contained in the genome. One of these rules may be employed by the visual system, where connections are formed on the basis of the interplay of molecular gradients and activity-dependent synaptic plasticity. We show how a simple model that accounts for such interplay can create both neural topographic maps and more complex patterns of ocular dominance, that is, the segregated binary mixture of projections from two eyes converging in the same visual area. With regard to the ocular dominance patterns, we show that pattern orientation may be instructed by the direction of the gradients of molecular labels. We also show that the periodicity of ocular dominance patterns may result from the interplay of the effects of molecular gradients and correlated neural activity. Overall, we propose that simple mechanisms can account for the formation of apparently complex features of neuronal connections.

Keywords: chemoaffinity; neural development; genome

The basics of information processing in the nervous system

The basic unit of computation in the brain is the nerve cell or neuron.[1,2] Like other cells in our bodies, neurons have a nucleus, a cell body, a cellular membrane, and can synthetize proteins and metabolize energy. Unlike other cells, neurons can also process information propagating through the neural net. Information processing depends on two types of protrusions formed by the neuronal cellular membrane. These protrusions are essentially small-diameter tubes (wires) that can extend large distances, branch, and form synapses with other neurons in the network. Two types of protrusions—dendrites and axons—play essential and different roles in information processing.

From a physical standpoint, a neuron is a capacitor, with two electrolytic plates formed by the ionic solutions inside and outside of the cells isolated from each other by a fatty cellular membrane. The capacitor is held at a constant voltage of about 60 mV by batteries in the form of ion pumps inside the membrane that actively transport ions across

the membrane.[3] When a nerve pulse arrives at a synapse, it changes the membrane voltage for a brief moment. This change in voltage can propagate along dendrites, which act, to the first approximation, as passive waveguides, toward the cell body. At the cell body, the voltage changes that are evoked by different synapses are combined (Fig. 1). When multiple synaptic events produce a combined large effect at the cell body, the neuron decides to emit a nerve pulse, also called an action potential or spike. A spike is a brief event, lasting a few milliseconds, during which the membrane voltage near the cell body drops close to zero or even gets reversed. Spikes are produced by nonlinear conductances in the membrane and are the basic packets that carry information across neural networks. Neurons in our cortex emit spikes several times a second. Spikes carry representations of our percepts, thoughts, feelings, and short-term memories. Spikes can be carried away from the cell body by axons. A cell usually forms a single axon that can branch and form synapses with dendrites of other cells in the network. Spikes reaching a synapse on an axon can

doi: 10.1111/nyas.12324

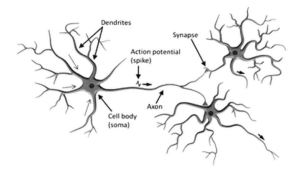

Figure 1. Basic elements of neural circuits. Information travels from synapses to cell bodies through dendrites and is combined at the cell body, which makes a decision to emit a spike. Spikes are usually emitted several times in a second. Spikes propagate along axons at high speed and reach synapses that have been formed with the dendrites of other cells. At the synapses, spikes can influence the dendrites of postsynaptic cells, leading to information transfer across the nervous system.

evoke a voltage fluctuation in the postsynaptic cell. This is how information spreads through the neural network.

The way in which information is processed in the brain depends on the synaptic–connectivity matrix. Our cortex contains about 10 billion neurons. Each neuron makes about 10^4 synapses. The total number of synapses in the human cortex is therefore close to 10^{14}. The formation of cortical connectivity is a task of enormous complexity. Although a substantial degree of neuronal connectivity is hard-wired (i.e., specified genetically), connections are also sensitive to experience, especially experiences presented during special periods of early neural development, which are called critical periods. According to the current dogma in neuroscience, although short-term memories spanning periods of a few seconds or hours are carried in spikes, the storage of long-term memories involves changes in the strengths of connections between neurons.[4–6] In neural networks, therefore, the "software," carried in the form of spikes, can modify network "hardware" (i.e., synapses), thus implementing the process of learning. The interplay between genetic factors and experience-dependent synaptic plasticity is one of the fundamental problems in neuroscience. Here, we will propose that genetic and experience-dependent factors can be combined within a simple mathematical description.

What is the information capacity of cortical connectivity?

The amount of information contained in the connections between neurons forms the upper bound on the amount of information that can be stored in long-term memory, assuming that it is stored in synapses.[4–6] To evaluate this quantity, we will use the formula from statistical physics:

$$I = \log_2 \Gamma. \qquad (1)$$

Here, I is entropy, which, in this case, represents information, while Γ is the number of possible combinations available in the system. For the string of N bits, for example, the number of combinations is $\Gamma = 2^N$ and Eq. 1 produces a sensible result of $I = N$ bits. To evaluate the amount of information stored in cortical connections, one has to calculate the number of connectivities possible between cortical neurons. For the human cortex, which includes $N \sim 10^{10}$ neurons and $s \sim 10^4$ synapses per neuron, the number of possible topological connectivities is $\Gamma \sim N^{Ns}$. This is based on the assumption that each neuron determines placement of its s synapses on N other neurons, leading to N^s combinations per neuron and $\Gamma \sim N^{Ns}$ combinations for the whole network. The amount of information that can be stored in the topology of the network is therefore close to the following:

$$I = Ns \log_2 N \sim 400 \text{ terabytes.} \qquad (2)$$

To put this number into perspective, let us assume that we record our visual experiences as a high-definition video using about 1 GB to store 1 h of continuous stream. Our cortical network allows us to record about 400,000 h of such a video, which amounts to about 45 years of continuous recording! This excludes, obviously, periods when visual stimuli are not available, that is, during sleep. Overall, a crude estimate of the amount of information available in just topological (structural) connectivity suggests that our cortical network is capable of storing much of our experiences throughout our lifetime.

The estimate presented above has several caveats. First, we assumed that synaptic strength is all-or-nothing. Some information can be stored in the strengths of individual synapses. Although some studies indicate that synapses may indeed be binary,[7–10] synaptic strength is likely a continuous variable.[4,11] Some simple estimates show, however, that the amount of information stored in the synaptic strengths is substantially smaller than the information capacity of the network topology.[12] Briefly, in both cases, the amount of information is proportional to the total number of synapses, as in Eq. 2. For topological connectivity, the coefficient of proportionality is $\log_2 N$, which is determined by the number of neurons where this synapse can be placed. The information capacity of continuous synapses is determined by the number of levels of synaptic strengths that can be measured owing to the finite signal-to-noise ratio (SNR) for each synapse h. The total information capacity of the neural network with continuous synapses is

$$I = Ns \log_2 N + Ns \log_2 h. \qquad (3)$$

Because synaptic SNR is not very big,[13] $h \ll N \sim 10^{10}$, the continuous nature of synaptic strengths determined by the last term in Eq. 3 is substantially smaller than the contribution from topological connectivity. For example, if synaptic SNR is about 10, then the last term is only 40 TB, which allows only 4.5 years of continuous video storage in our simplified example. Though this is an enormous contribution, it pales in comparison to the 45 years of video stream permitted by the topological connectivity given by Eq. 2.

The second questionable assumption in the estimate from Eq. 2 is that a neuron can potentially make synapses to all other neurons in the network. In fact, synapses are more often made locally to neurons within a small neighborhood.[14,15] This argument does not substantially change our estimate, but allows us to refine it somewhat. Indeed, let us assume that the network is broken into small chunks of n neurons each. In the language of neuroanatomy, these chunks are often called cortical columns.[16] Consider a cortical neuron that makes synapses with other cortical neurons. This neuron will first decide which columns it needs to send its axon to. Once the axon arrives at the column, it will, to the first approximation, make synapses in and around that column. The number of combinations in the connectivity constrained by these considerations is $\Gamma \sim \left(\frac{N}{n} n^s \right)^N$, yielding the topological information of

$$I \approx Ns \log_2 n. \qquad (4)$$

Some arguments suggest that the size of the cortical column is close to $n \sim \sqrt{N} \sim 10^5$,[14,15] reducing by a factor of two our simple estimate (Eq. 2). If axons are allowed to branch, topological information is increased in comparison to Eq. 4, so the real value is expected to be somewhere between Eqs. 2 and 4. These estimates suggest that cortical connectivity is capable of storing the amount of information equivalent to perhaps several decades of visual experience.

An important question is whether cortical networks can gain access to this combinatorial complexity. An access to this information capacity would require that topology of neural network be remodeled to accommodate ongoing experience throughout lifetime. Although in some brain areas, such as the rodent somatosensory cortex, synaptic remodeling is indeed observed in normal conditions throughout the life span[17–21] in others, new synapses are created when something unusual happens, such as retinal lesions.[20] It is not clear, therefore, whether neural networks can readily change their topology to reflect ongoing experience in the adult brain. Therefore, cortical networks may have to resort to changes in the strengths of existing synapses, described by the second term in Eq. 3, to store information during adult life.[22] Changes in synaptic strength represent substantial information capacity, which is not, however, as large as the topological information. In most cortical areas, changes in synaptic topology may be restricted to specialized stages of neural development early in the organism's life known as critical periods. It is interesting, therefore, to understand the rules by which connections are formed during neural development.

If topological synaptic connectivity is established, at least in part, during early neural development, then an interesting discrepancy can be noticed between the amounts of information in connectivity (Eq. 2) and in the genome. Because the haploid human genome contains only about 1 GB of information (about 3 billion base pairs), the question is, how is it possible to set up 400 TB or so of connections with so little information? It appears that, by analogy, a 1-h instructional video contained in the genome is sufficient to explain how to make 400,000 similar videos contained in the topology of cortical connectivity. In addition, the genome has to accomplish much more than just instructing the creation of the neural network.

The discrepancy between genomic and neural information content can be understood if one remembers that they represent two different types of information. Genomic information describes the complexity of neural connectivity in the Kolmogorov sense, that is, the length of the algorithm needed to set it up.[23] Such algorithms can be very short. For example, one could be satisfied with a simple instruction that connectivity is established completely randomly. Such an algorithm can be very short and sufficient to set up many terabytes of connectivity. What this discrepancy really means is that the position of every synapse cannot be specified in the genome. Of course, it is hard to imagine that the topology of a neural network established during neural development is fully random, especially if it cannot be easily changed later in life. Perhaps, more sensible simplifying rules are used to set up a vast number of connections on the basis of the limited information contained in the genome.[24] Here, we will describe one of these simplifying rules that relies on the existence of neural maps.

Maps in the brain: an evolutionary adaptation that reduces information necessary to set up connectivity

Connectivity in the cortex and beyond can be characterized by various maps that describe how features of the outside world are represented in the brain.[25] One well-known example of these maps is the somatosensory homunculus: an ordered representation of the surface of our body on the two-dimensional (2D) surface of the cortex. In the visual system, a widespread principle of organization is the retinotopic or topographic map.[16] Within this map, the surface of the retina is reflected topographically on the 2D surface of various visual brain regions. This organization is, in principle, analogous to the somatosensory homunculus, because topographic maps represent the sensory surface of the retina on the surface of the brain. Topographic maps are produced by connections between various brain regions that preserve neighborhood relationships between neurons. For example, retinal ganglion cells that cover the 2D retinal surface send their axons to a brain region called the superior colliculus. In the superior colliculus, axons spread out and form an ordered representation of the 2D retinal surface. This means that two neighboring retinal neurons

send their axons that terminate near each other in the target brain region.

Most visual cortical and subcortical areas display some form of topographic organization. For example, another retinal axon recipient region known as the lateral geniculate nucleus (LGN) also displays topographic organization.[16,27] Furthermore, the second-, third-, and higher order visual areas are also organized topographically, by virtue of inheriting the ordering from lower areas.[16,28] Thus, neurons from the LGN send their axons to the primary visual cortex (V1), which preserves topographic organization leading to the continuous representation of 2D visual space in V1.[27] Topographic ordering seems to be a general feature of visual maps that span multiple levels of organization.[16]

The functional significance of topographic maps is not clearly understood. It was suggested that topographic projection helps minimize wiring length between neurons in the target needed to process visual information. Such wiring is expected to connect neurons in the target that respond to neighboring points in the image.[25] Here, we propose an alternative explanation that is based on the need to reduce space in the genome, which in turn is needed in order to define connectivity between the retina and its targets. Our explanation relies on the mechanisms by which topographic connectivity is formed, which will be discussed in some detail later.

These mechanisms have been under investigation for many decades. It was discovered that the axons of retinal cells can recognize and respond to molecular tags that are expressed (produced) by cells in the target[29–31] (Fig. 2). Because the retinal surface is 2D, two sets of molecular tags exist that serve as molecular guidance cues for vertical and horizontal retinal directions. For the horizontal retinal axis (the *X*-axis in Fig. 2), the molecules called EphA receptors and their ligands ephrin-As are thought to implement topographic ordering.[29–31] Axons of retinal ganglion cells that carry EphA receptors can sense the level of ephrin-A in the target. From the layout of maps in Figure 2A, it can be hypothesized that high levels of ephrin-A repel EphA-carrying axons. This suggestion is supported by experimental evidence.[29–31]

Similarly, the vertical coordinate in the retina is defined by the levels of EphB receptor (Fig. 2B). The recipient coordinate is determined by the levels of expression of ephrin-B in the target. From the layout

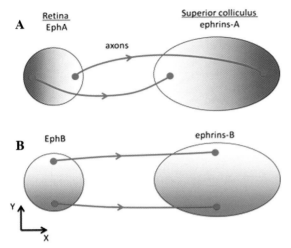

Figure 2. Two systems of molecular gradients responsible for topographic map formation between the retina and superior colliculus. The horizontal axis in the retina is encoded by the level (darker = higher) of EphA receptor, which is a surface-bound protein capable of binding and recognizing its ligand, ephrin-A, on surrounding cells. (A) The corresponding axis in the target is set up by the level of expression of ephrin-A in the target. (B) The vertical axis in the retina is encoded by another surface-bound receptor EphB. The combined levels of the expression of receptors of the EphB subfamily form an overall gradient from low on top to high at the bottom. The corresponding axis in the target is set up by the levels of ephrin-B, which serves as a ligand for EphB.[26,34]

of the map (Fig. 2B), it could be hypothesized that the interaction between EphB-carrying axons and ephrin-B can be described as chemoattraction,[29–31] although alternative mechanisms have also been proposed.[32,33] Thus, the mapping of two retinal axes onto two dimensions in the brain is defined by two approximately independent systems of gradients: EphA/ephrin-A and EphB/ephrin-B.[29–31,34–38]

To describe the formation of this projection, several quantitative models have been proposed.[34] Here, we will present a model that is based on the simple Hamiltonian. To describe the interaction between EphA receptors and their ligands (i.e., ephrin-As), one can assume that the following functional is minimized:

$$H = \alpha \sum_i q_i^A \varphi^A(\vec{r}_i). \qquad (5)$$

Here, q_i^A is the level of EphA expression by an axon number i, $\varphi^A(\vec{r}_i)$ is the level of ephrin-A ligand at point $\vec{r}_i$, where this axon terminates, and the

sum is assumed over all axons in the retina. This functional represents the total number of molecules of EphA receptor bound by ephrin-A ligand within a mass–action law, with a coefficient of proportionality determined by the binding constant. When the Hamiltonian is minimized, the layout of axons in the target $\vec{r}_i$ minimizes the total number of bound EphA/ephrin-A pairs, thus implementing the chemorepulsion in the interaction of this pair of molecules (Fig. 2A). As highlighted by our notations, there is also an analogy between this system and a set of charges q_i^A moving in the external electric field defined by the electric potential $\varphi^A(\vec{r}_i)$.

After examining the Hamiltonian (Eq. 5), one naturally wonders why all neurons do not project to the same point, where the level of ligand $\varphi^A(\vec{r}_i)$ is the minimum. The definitive answer to this question is not yet available.[34] Some other factor should be involved to distribute axonal termination positions $\vec{r}_i$ uniformly throughout the target. Some experimental data[34] are consistent with the model presented here (Fig. 3) that competition between axons for space in the target prevents them from all terminating at the minimum of $\varphi^A(\vec{r}_i)$. This model follows a long trail of studies suggesting that competition between axons in space makes the density of projections in the target approximately uniform.[34] This means that the terminations of axons, that is, the places where most synapses are made, have roughly the same density in the target. The terminations of axons are symbolically shown in Figure 3 by circles of finite diameter, which prevents them from occupying the same positions in space. In practice, we can enforce this constraint by assuming that the axonal termination position $\vec{r}_i$ can occupy positions limited to a square grid, with a single axon per grid point. The Hamiltonian can then be minimized by shuffling axonal termination positions in the target.

This approach can also be used to include topographic mapping along the vertical retinal axis that is dependent on EphB–ephrin-B interactions (Fig. 2B). To describe these interactions, we add the second term to the Hamiltonian:

$$H = \alpha \sum_i q_i^A \varphi^A(\vec{r}_i) - \beta \sum_i q_i^B \varphi^B(\vec{r}_i). \qquad (6)$$

Because the interactions between axons carrying EphB and the ephrin-B environment can be described as attraction (Fig. 2B; however, see Refs. 33 and 34), the overall sign in front of the second term

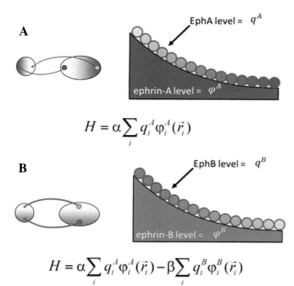

A

$$H = \alpha \sum_i q_i^A \varphi_i^A(\vec{r}_i)$$

B

$$H = \alpha \sum_i q_i^A \varphi_i^A(\vec{r}_i) - \beta \sum_i q_i^B \varphi_i^B(\vec{r}_i)$$

Figure 3. Interactions (A) between EphA receptors and ephrin-A ligands, as well as (B) between EphB receptors and ephrin-B ligands, can be described by a simple Hamiltonian. Each axonal termination location (circle) carries a charge determined by the level of the receptor expressed by the neuron. Because there are two types of receptors, EphA and EphB, axons carry two types of charges, q_i^A and q_i^B, carried by each axon number i. The charges interact with two fields determined by the levels of ligand in the target at the position of axonal termination, $\varphi^A(\vec{r}_i)$ and $\varphi^B(\vec{r}_i)$. These fields are analogous to electric potentials or gravitational potentials for a regular particle. That axons occupy finite space is represented by their finite size. When the Hamiltonian is minimized, space exclusion (i.e., competition for space) makes axons with different levels of receptor spread around the target; A shows the Hamiltonian for EphA only, while B contains Hamiltonians for both sets of molecules.

is negative. According to this Hamiltonian, therefore, axons would like to increase the amount of EphB bound by ephrin-B. As illustrated in Figure 2B, minimization of this Hamiltonian, combined with the competition between axon terminals for space, can implement the formation of topography in a 2D projection.

Here, we introduced the mechanism of topographic map formation based on molecular gradients. This mechanism uses a small number of molecules to establish a large number of connections. Topographic projection between the retina and superior colliculus includes about 10^6 axons and can be established with only two pairs of molecules. This mechanism can therefore be defined in a limited space in the genome. Thus, to-

pographic maps and the underlying molecular gradients can be viewed as evolutionary adaptations aiming to limit the information used in the genome to set up connections, as we described earlier. Because similar mechanisms are reused throughout the visual system, including dependence on the same set of molecules,[27,39] a large number of neural connections can be established through a similar model.

A more general form of chemoaffinity

To make Eq. 6 more compact, one can introduce an interaction matrix between labels $M_{\mu\nu}$, where indexes μ and ν can take values A and B. For the case of a topographic map, $M_{\mu\nu}$ is a two-by-two diagonal matrix describing the affinities between A and B molecules:

$$\hat{M} = \begin{pmatrix} \alpha & 0 \\ 0 & -\beta \end{pmatrix}. \tag{7}$$

Equation 6 can then be rewritten in the following form:

$$H = \sum_{\mu,\nu=A,B} M_{\mu\nu} \sum_i q_i^\mu \varphi^\nu(\vec{r}_i). \tag{8}$$

Thus every axon can be viewed as carrying two types of charges, each interacting with a diverse set of fields $\varphi^\nu(\vec{r}_i)$. The strength of this interaction is determined by matrix $M_{\mu\nu}$.

More generally, in case of several labels, the sum in Eq. 8 can be extended to include all these labels. In this case, $M_{\mu\nu}$ does not have to be square and diagonal. To generalize the Hamiltonian (Eq. 8) even more, we will assume that an axon can make several connections at different locations. If index i describes an axon and index j enumerates dendrites of different cells, the weight matrix W_{ji} defines the strength of connection between axon i and dendrite j. The Hamiltonian (Eq. 8) can then be rewritten as

$$H = \sum_{\mu,\nu=A,B} M_{\mu\nu} \sum_{ij} \varphi_j^\nu W_{ji} q_i^\mu. \tag{9}$$

Equation 8 can be obtained from Eq. 9 by assuming that connections are made between a neuron number i terminating at point $\vec{r}_i$ and a dendrite at the position $\vec{r}_j$, that is, $W_{ji} = \delta_{\vec{r}_j,\vec{r}_i}$. The goal of neural development is then in converting the interaction matrix $M_{\mu\nu}$ into the connection matrix W_{ji}

by minimizing the Hamiltonian (Eq. 9). Because the interaction matrix is defined genetically, this mechanism allows the transfer of genetic information into the topology of neural circuits.

How to combine the effects of neural activity and molecular labels

So far, we have discussed the impact of molecular tags on the structure of neural connectivity. The interactions between these molecules and their distribution in space are, to the first approximation, defined genetically. To describe these molecular interactions within a mathematical model, we used an approach that was based on the minimization of a Hamiltonian, such as defined in Eqs. 6 and 9. Our next goal is to include the effects of experience into the model. Because organism's sensory experience is initially represented in the nervous system in the form of neural activity, that is, spikes, understanding how activity affects connections may yield clues into the mechanisms of learning. When effects of genes represented by molecular labels are combined with the activity-dependent learning within the same model, one has a chance to understand the interplay of nurture and nature in the network topology. One way to account for the effects of neural activity, that is, spikes, is to add an extra term to the Hamiltonian discussed earlier.[36–38] To motivate the form of this extra term, we will examine the results of experiments in which the activity of retinal cells is changed or eliminated.

What generates activity in the visual system when topographic connections are formed? In mice, the formation of topographic maps occurs in young pups, when the eyes are still shut. The activity of retinal ganglion cells in these animals is determined by the retina itself, not visual stimuli. During this period, the retina emits spontaneous waves of activity that emerge periodically and propagate through the retinal ganglion cell layer.[34,40–42] These spontaneous waves have a wavelength λ roughly equal to 1/10 of the size of the retina. They serve as training stimuli that prepare the retina for normal operation when the animal's eyes are open. The pair-wise correlation function between retinal cells due to the waves can be monitored by using arrays of electrodes.[34,40] It can be approximated[34,36–38] as

$$C_{ij} = \exp\left(-|\vec{R}_i - \vec{R}_j|/\lambda\right). \qquad (10)$$

Here $\vec{R}_i$ and $\vec{R}_j$ are positions of cell bodies (axonal points of origin) in the retina, not in the target. A very similar pattern of correlations between retinal cells is observed in tadpoles[43] during the period relevant to topographic map formation. In contrast to mice, activity in tadpoles is produced by visual stimuli. In both mice and tadpoles, correlated activity affects retinal projections similarly, as described later. Thus, although biological origins of activity are different in different species, the effects of activity correlations could be similar. Understanding these effects may help formulate a mathematical model for activity-dependent learning in neutral networks.

To understand the effects of retinal activity, it is useful to look at experiments in which it is changed. Retinal waves are disrupted in mutant mice in which neurotransmission between retinal cells is perturbed, called $\beta 2^{-/-}$ knockout mice.[40] The impact of the disruption of retinal waves on topographic connectivity is illustrated in Figure 4. Connections between the retina and its target can be studied by tracing axons using fluorescent dyes. If a tracer is injected into a small area of the retina, it can diffuse along the axons and emerge in the target at the points where axons terminate. This method allows the monitoring of axonal trajectories as they navigate their way to their destination. An injection of tracer in a small point of the retina of a normal animal leads to the appearance of a small dot in the target (Fig. 4A), indicating that topographic maps are usually quite precise.[40,44] In $\beta 2^{-/-}$ mice, a tracer injection shows that axons are distributed over a large area (Fig. 4B), suggesting that the map is less precise in these animals.[40,44] This finding indicates that the role of retinal waves and retinal activity in general is to make the topographic map more precise. This result can also be explained by suggesting that the axons that leave the retina from neighboring points are attracted to each other in the target. In normal (wild-type) animals (Fig. 4A), the axons are strongly attracted to each other, leading to their convergence to a focal point in the target. In the mutant animals (Fig. 4B), the attraction is not as strong; thus, the axons do not condense to a small point and the topographic map is not so precise.

To describe pair-wise attraction between axons in the target, we derived the additional term to the Hamiltonian that is minimized in the conventional

Normal retinal waves

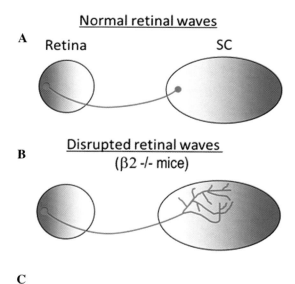

A
Retina   SC

B
Disrupted retinal waves
(β2 -/- mice)

C

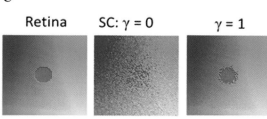

Retina   SC: γ = 0   γ = 1

Figure 4. The effects of correlated neural activity (spikes) on topographic projection. The activity of retinal axons results from activity waves propagating through the retina. These waves are generated by the retina itself to entrain axons projecting to retinal targets. (A) In normal mice, a small patch of neurons (red circle) sends axons to a compact area in the target. (B) In mice with disrupted retinal waves ($\beta2^{-/-}$ knockouts), these axons project over a large distributed region.[40] (C) The topographic map is shown by the color map. The projection of a small circle of axons at the center of the retina is indicated by red points. The findings illustrated in A and B can be explained by assuming that axons with correlated activity, that is, located near each other in the retina, are attracted to each other (see text). In $\beta2^{-/-}$ mice ($\gamma = 0$), this attraction is weak. The weak remaining ordering is due to Ephs interacting with ephrins. In the case of strong attraction ($\gamma = 1$), due to a strong attraction between axons emanating from a small circle, these axons cluster together in the retina, leading to an almost precise image of the circle and a more precise topographic map.

Hebbian learning theory.[36] This activity-induced Hamiltonian has the following form:

$$H_{\text{act}} = -\frac{\gamma}{2} \sum_{ij} C_{ij} U\left(\vec{r}_i - \vec{r}_j\right). \tag{11}$$

This Hamiltonian defines pair-wise interactions between axons. The overall strength of this interaction is determined by the parameter γ. This parameter can be used to manipulate the strength of activity-dependent contributions and can be chosen from best fits to experimental data. The strength of interaction between pairs of axons is determined by the correlations in their activity C_{ij}, given by Eq. 10. Two axons are attracted to each other with the distance-dependent interaction potential.[36,38,45]

$$U(r) = \exp(-r^2/2a^2). \tag{12}$$

To put this model into perspective, we note here that the pair-wise interactions described by Eq. 11 are similar to the electrostatic interaction energy described by Coulomb's law:

$$H_{\text{electr}} = \frac{k}{2} \sum_{i \neq j} q_i \cdot q_j \frac{1}{\left|\vec{r}_i - \vec{r}_j\right|}. \tag{13}$$

Axon–axon interaction potential is equivalent to the Coulomb potential, $U(\vec{r}_i - \vec{r}_j) \leftrightarrow -1/\left|\vec{r}_i - \vec{r}_j\right|$, while the axon–axon activity correlation is similar to the product of two electric changes $C_{ij} \leftrightarrow q_i \cdot q_j$. Interestingly, the neural network Hamiltonian (Eq. 11) is both similar to and more complex than electrostatic energy (Eq. 13), because the strength of interactions between two particles C_{ij} cannot be represented as the product of two charges. Instead, it depends on the distance between the two retinal cells. This observation leads to some inconvenience because there is no straightforward method to introduce an analog of electrostatic potential $\phi(\vec{r}) = k \sum_j q_j / \left|\vec{r} - \vec{r}_j\right|$. Equation 11 is a more general form of interaction than that which is used in physics.

The activity-dependent Hamiltonian (Eq. 11) was derived assuming that it implements learning in neural networks.[36] It favors close positioning of neurons with correlated activity. This property implements Hebbian learning rules, according to which neurons that carry correlated activity form strong connections. The attractive Hamiltonian (Eq. 11) can be understood as implementing this rule, because it makes coactive neurons more likely to be connected by placing their synapses in close proximity.

By adding this activity-dependent Hamiltonian (Eq. 11) to Eq. 8, we obtain the total Hamiltonian that will be used in the remainder of this paper.

$$H = \sum_{\mu,\nu=A,B} M_{\mu\nu} \sum_i q_i^\mu \varphi_i^\nu(\vec{r}_i)$$

$$-\frac{\gamma}{2}\sum_{ij} C_{ij} U\left(\vec{r}_i - \vec{r}_j\right). \qquad (14)$$

This Hamiltonian can describe a large array of different experiments.[33–38] Here we will show that it can also explain the formation of higher order maps, such as ocular dominance (OD) columns.

Maps of OD

Another example of ordered connectivity in the brain is the map of OD.[46–48] Neurons in the primary visual cortex of higher mammals are dominated by one of the eyes. The binary mixture of two types of neurons, right- and left-eye dominated, segregates during neural development into the set of domains formed on the 2D surface of cortex[46–48] (Fig. 5A). The layout of these domains on the cortical sheet is known as the ocular dominance pattern (ODP). The most frequent structure of ODP resembles zebra skin: the left- and right-eye–dominated neurons form interdigitating stripes, while other patterns are also possible, such as that resembling leopard skin.[25,49,50] The width of OD stripes can be affected by manipulations with neural activity.[51] Also, ODP orientation follows systematic trends preserved across several species. Many theories have been proposed to explain various features of ODPs.[25,49,52–56] Here, we will demonstrate that the simple Hamiltonian introduced earlier, with minor modifications, can provide the mechanism of ODP formation. Our approach also enables us to calculate the pattern periodicity and stripe orientations as well as other interesting features about ODPs.

One interesting system where ODPs are observed is the tectum of three-eyed frogs.[57,58] The tectum in frogs is an area homologous to the mammalian superior colliculus described earlier. Axons of retinal cells reach the tectum and establish topographic maps there, similarly to the mouse. Naturally, animals have two tecta, one in each half of the brain. In frogs, the axons from each of the two eyes completely cross to the opposite tectum, leading to two halves of the brain that are fully dominated by one of the eyes (Fig. 5B). If the third eye is introduced surgically, this extra eye can grow connections to one of the tecta and establish a topographic map there. The

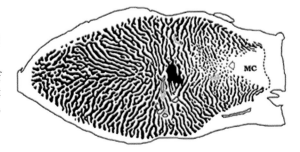

Figure 5. Ocular dominance patterns in various systems. (A) Macaque monkey primary visual cortex (left).[47] The interdigitating domains dominated by left- and right-eye dominance form stripe patterns, with the exception of patchy maps in the proximity of the monocular crescent (MC). (B) In normal frogs, the axons from both eyes almost fully cross to the other side, so that each hemibrain is dominated by either the left or the right eye. (C) In frogs in which a third ectopic eye was surgically introduced, the projections from a normal eye (#1), and the ectopic eye (#3) form interdigitating stripes resembling those in the primate visual system (A).[57,58]

axons of this extra eye establish connections within domains resembling stripes (Fig. 5C) similar to the ODPs observed in primates (Fig. 5A). Interestingly, axons in the visual system of such three-eyed frogs face a novel problem of establishing connections based on their eye (ocular) dominance. The solution that these axons find is similar to the solution observed in higher mammals (Fig. 5A). When axons from two eyes segregate into distinct domains, the cells in the target become dominated by one of the eyes, leading to the emergence of ODP. That axons segregate into similar patterns in different animals suggests that the mechanisms of the formation of ODP are similar.

The conventional mechanism for ODP formation described in the theoretical literature appeals to the mechanisms similar to Turing instability,[53–56,59,60] which suggests that axons interact with each other through a pair-wise potential that has a negative (divergent) Fourier harmonic. As a result, the uniform binary mixture of two types of axons, arriving from the two eyes, is unstable. The corollary of this suggestion is that the spatial periodicity of this Fourier harmonic that determines ODP periodicity is not easy to change.[51] Experimental evidence shows that ODP period can change in a wide range through pharmacological manipulations.[51] Motivated by this observation, we will introduce here a different mechanism within which ODPs are generated by molecular gradients.

The mechanism is based on the Hamiltonian introduced earlier. Indeed, let us assume that axons of two eyes (i.e., eyes #1 and #3) of the three-eyed frog are molecularly indistinguishable. This means that they carry the same distributions of molecular labels EphA and EphB. When these two groups of axons are placed into the same map, if only molecular labels contribute to the OD map formation, they should form a uniform mixture of two eyes that will be called here the salt-and-pepper (S&P) configuration (see Refs. 25, 49, and 50 for definition of this pattern). This is because if molecular labels are identical between eyes, then they cannot be distinguished on the basis of molecular labels alone. The full Hamiltonian for this system of axons, however, includes correlation in their activity (spikes). This correlation can distinguish two eyes and potentially contribute to ODP formation. Indeed, our Hamiltonian is

$$H = \sum_{\mu,\nu=A,B} M_{\mu\nu} \sum_i q_i^\mu \varphi^\nu(\vec{r}_i)$$
$$- \frac{\gamma}{2} \sum_{ij} Y_{d_i d_j} C_{ij} U\left(\vec{r}_i - \vec{r}_j\right). \quad (15)$$

Here, we used the same Hamiltonian as we introduced before in Eq. 14. The only difference is that we introduced the eye–eye correlation matrix $Y_{d_i d_j}$. This matrix defines the strength of correlations in the activity of two axons i and j depending on their affiliation with the eyes d_i and d_j. Matrix $Y_{d_i d_j}$ is therefore a two-by-two matrix. If eyes are similar in their parameters, one can expect this matrix to be symmetrical. One can also expect that the same eye correlations ($d_i = d_j$) described by the diagonal elements of $\hat{Y}$ will be greater than different eye cor-

relations ($d_i \neq d_j$), that is, $Y_{1,1}, Y_{2,2} > Y_{1,2}, Y_{2,1}$. In the case of correlations in activity introduced by retinal waves, this is natural because retinal waves in different eyes are only weakly correlated.[43] For the activity induced by visual experience, such as in frogs, the cross-eye correlations are weaker due to incomplete convergence of eyes on the same stimulus. We will use the eye–eye in the form

$$\hat{Y} = \begin{pmatrix} 1 & 0 \\ 0 & 1 \end{pmatrix}. \quad (16)$$

When this Hamiltonian is used to find the arrangement of retinal axons from two eyes, such as in the three-eyed frog, the following projection is observed (Fig. 6).

Figure 6A and B shows the OD map resulting from the minimization of our Hamiltonian. These maps result from the interplay of two conflicting factors. On the one hand, chemical gradients that do not distinguish between two eyes have a tendency to build perfect topographic maps by mixing the two eyes uniformly, that is, favoring the S&P configuration. Chemical labels therefore favor ODP with a very small (zero) period. Correlations in neural activity, however, through the eye–eye correlation matrix $\hat{Y}$, introduce attraction between axons of the same eye. These correlations therefore favor the close positioning of axons belonging to the same eye that fill the same-eye domains. The finite periodicity of the ODPs results from the interplay of these conflicting factors.

ODP as a result of competition between molecular gradient and correlated activity

To make this argument quantitative, here we derive the periodicity of ODP. Our qualitative arguments from the previous section suggest that when correlated activity is weak ($\gamma \to 0$), ODP period should be determined by the molecular gradients, and, therefore, become small. Conversely, with strengthening of the effects of correlated activity ($\gamma \to \infty$), ODP period should diverge indefinitely. Our goal in this section is to confirm this argument by detailed calculation.

Using the last Hamiltonian (Eq. 15), we will calculate now the change in the variable H/N (energy per neuron) due to the formation of ODP. We will assume that ODP is formed from a completely mixed state in which left- and right-eye neurons are

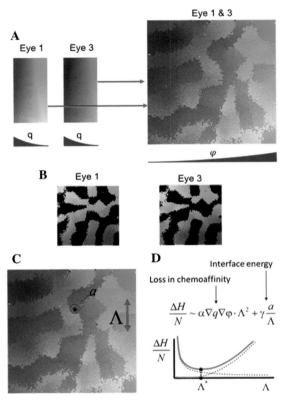

Figure 6. The results of numerical minimization of our Hamiltonian (Eq. 15) for the case of the three-eyed frog. (A, B) We used identical chemical labels for two eyes as before.[36,38] Two identical rectangular arrays of 50 × 100 neurons representing two eyes project into a single tectum described by a 100 × 100 square. Activity of two eyes was not correlated, according to Eq. 16. Lack of cross-eye correlations leads to the formation of ODP superimposed on the topographic map. (C) The periodicity of the pattern is determined by the interplay between the chemoaffinity and interface energy, as described in the text. Neurons within the disk of radius a have a reduced activity-dependent contribution leading to the emergence of interface energy (Eq. 19), shown also in D by the dotted line. (D) Interplay between the molecular labels and activity-dependent contributions (two dotted lines in the plot) defines the optimal periodicity of ODP.

intermingled. This configuration, an S&P pattern, can be viewed as an ODP pattern of very small periodicity. The simplest to understand is the change in the first term of the Hamiltonian that determines the energy due to molecular labels (by the term *energy*, we understand the values described symbolically by Eq. 15 rather than the real physical energy of the chemical bonds). Because the equilibrium for molecular labels is a fully mixed S&P pattern (see discussion after Eq. 16), when neurons are re-

arranged into ODP, they are moved a distance Λ from their equilibrium positions. Here, Λ is the stripe width. Change in energy due to molecules (Ephs) per neuron is expected to be $\Delta H_{\mathrm{mol}} = H_{\mathrm{ODP}} - H_{\mathrm{S\&P}} \sim N\alpha \cdot |\nabla\varphi\nabla q| \cdot \Lambda^2$, which is just a Hooke's law, that is, the energy of the elastic "tether" connecting an axon to its equilibrium position (Fig. 6C). Because there are two labels, EphA and EphB, and due to different possible orientations of stripes with respect to the gradients of receptors $\vec{\nabla}q$ and ligands $\vec{\nabla}\varphi$, additional care should be taken if this term is to be computed exactly. Simple calculations show that indeed the exact form of the molecular change in energy is

$$\frac{\Delta H_{\mathrm{mol}}}{N} = G\Lambda^2, \tag{17}$$

where

$$
\begin{aligned}
G(\vec{n}) &= \frac{1}{24}\left[-\alpha(\vec{\nabla}\rho^A\vec{n})(\vec{\nabla}\varphi^A\vec{n}) + \beta(\vec{\nabla}\rho^B\vec{n})(\vec{\nabla}\varphi^B\vec{n})\right] \\
&= \frac{1}{24}\sum_{\mu\nu} M_{\mu\nu}(\vec{\nabla}\rho^\mu\vec{n})(\vec{\nabla}\varphi^\nu\vec{n}). \tag{18}
\end{aligned}
$$

Here, $\vec{n}$ is a unit vector perpendicular to the stripes. The coefficient $G(\vec{n})$ describes the effects of molecular gradients on ODP. It is expected to be positive because the gradients of the receptor and ligand should be anticorrelated for a repulsive label (EphA) and correlated for attractive labels (EphB).

On the other hand, the average energy change for each axon due to activity correlations between axons is

$$\Delta H_{\mathrm{act}} = -\frac{\gamma}{2}\pi(\sigma a)^2 + 2\sqrt{2\pi}\gamma(\sigma a)^2\frac{a}{\Lambda}. \tag{19}$$

In deriving this equation we assumed that the range of activity-dependent interactions a is much smaller than ODP period Λ, that is, $a \ll \Lambda$. Our subsequent results will be exact in this asymptotic limit. In Eq. 19, σ is the mean 2D density of axons in the cortex. The first term in Eq. 19 is the condensation energy: the gain in energy due to surrounding an axon with 100% of axons of the same eye, as compared to 50% in the S&P configuration. The second term is the interface energy, that is, the loss of energy by the axons at the interface between the two eyes. This loss is proportional to the fraction of axons at the interface a/Λ, where a is the range of attraction defined in Eq. 12 (red disk in Fig. 6C). The interface axons lose a substantial part of their condensation energy (Fig. 6C).

By minimizing $\Delta H = \Delta H_{\text{mol}} + \Delta H_{\text{oct}}$ (Fig. 6C), we find the optimal width of the stripe:

$$\Lambda_{\min} = a \left(\frac{\sqrt{2\pi}\gamma\sigma^2}{G} \right)^{\frac{1}{3}}. \tag{20}$$

Equation 20 describes how the pattern periodicity changes as a function of strength activity correlations γ and the molecular gradients. According to this equation, the periodicity of ODP is determined by the interplay between the effect of gradients and correlated activity: the former tends to decrease the period toward the uniform mixture of eyes, while the latter increases the size of the domains. The periodicity of ODPs can therefore vary in a wide range.

Experimentally, the periodicity of ODP can be affected by manipulations of neural activity. Pharmacological enhancement of inhibition in cat area 17 can substantially increase ODP period.[51] Our model allows the interpretation of these experiments. To do so, we have to assume that, owing to a homeostatic mechanism, the increase in inhibition is accompanied by an increase in excitatory NMDA synaptic conductance.[61] This form of balance has been proposed as a mechanism that maintains constant levels of activity in the network.[62] Assuming homeostasis between excitation and inhibition, enhancing inhibition increases parameter γ of our model, leading to wider OD stripes, according to Eq. 20. Overall, we propose that the formation of ODPs can be explained by a simple topographic model as the interplay between the effects of molecular gradients and experience-dependent activity. An increase in the effects of activity during the critical period for ODP formation leads to an increase in ODP periodicity.

The orientation of ODPs can be instructed by molecular labels

Here, we will show how only a few molecules forming a set of gradients in the cortex can create the apparently complex features of cortical maps, such as the pattern of ODP orientation. As we showed earlier, ODPs can result from the balance of the effects of molecular gradients and correlations in neural activity. The former are determined genetically, while the latter can be dependent on organism's experience. In primates, OD stripes display an interesting pattern of orientation. In the area of the cortex corresponding to the center of the visual world, the stripes run horizontally in the real world, while at the periphery, the stripes are approximately concentric (Fig. 7B).[52,63] This remarkable phenomenon has been explained from the point of view of shortening the wires (axons and dendrites) that process information from two eyes in the cortex.[52] Although the evolutionary (teleological) significance of the stripes' orientation may be understood,[52] it is not clear mechanistically how the optimal pattern of orientations is formed during neural development. Here, we will argue that stripe orientation may result naturally from the orientation of two systems of interacting molecular gradients in the eyes and in the cortex.

To understand the factors determining stripe orientation, assume that one of the gradients, such as the gradient of B molecules, is weak or nonexistent. In this case, in 2D, the stripes will tend to orient themselves parallel to the remaining gradient of A molecules. Using Eqs. 17 and 19, it is easy to calculate the value of the Hamiltonian for the optimal wavelength:

$$\Delta H = -\frac{\gamma}{2}\pi\sigma^2 a^2 + 3\,(2\pi)^{\frac{1}{3}}\,a^2\sigma^{\frac{4}{3}}\gamma^{\frac{2}{3}} \cdot G(\vec{n})^{\frac{1}{3}}. \tag{21}$$

The important factor in this expression that depends on stripe orientation is $G(\vec{n})$ given by Eq. 18.

$G(\vec{n})$ should be minimized over orientations to minimize the Hamiltonian. To minimize G, in the case of a single system of molecules such as A, the vector normal to the stripes $\vec{n}$ has to rotate perpendicularly to the gradients of these molecules, which leads the stripes to be parallel to the gradients (Eq. 18). Because $G(\vec{n})$ is proportional to the gradients of both receptors and ligands, in the case of the receptor gradient $\vec{\nabla}q^A$ and ligand gradient $\vec{\nabla}\varphi^A$ pointing in different directions, to minimize G, the stripes orient themselves parallel to the median direction between them, as follows from a simple calculation. The situation is more complex when both systems of molecules, A and B, are of similar strengths. In this case, the stripes have to choose a preferred direction between A and B: if gradients of A molecules are weaker, the stripes will prefer to orient in the direction preferred by B. The interplay between two systems of molecular gradients may create the possibility for a sharp transition in OD stripe orientation, which is similar to what is

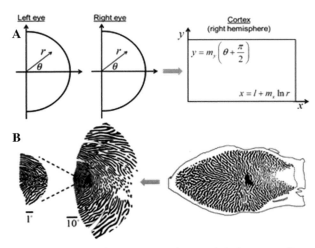

Figure 7. The geometry of topographic maps in the primate visual cortex. (A) The 2D coordinates in the cortex approximately represent polar retinal coordinates. The two halves of the left and right retina representing the same points in the visual world converge to the same cortical hemisphere. (B) When cortical OD stripes (right, adopted from Ref. 47) are projected back to the retinal space (left, adopted from Ref. 63), the orientation of stripes shows two trends. In the middle of the eye, stripes tend to run horizontally, while at the periphery, stripes are organized concentrically.[63]

observed experimentally. This intuition is confirmed by our detailed calculations later. To introduce our argument, we have to describe the geometry of the topographic map in the primate visual cortex, which we will do next.

In mice and lower vertebrates such as frogs, the map between retina and brain targets is approximately uniform (Fig. 4). This means that the magnification factor of the transformation, also known as its Jacobian, is roughly constant throughout the retinal space. By contrast, in primates as well as in other higher mammals, topographic maps encode a strongly nonuniform transformation. This transformation leads to an enormous overrepresentation of the retinal center in the brain. Our strategy for sampling the visual world may be simply to move the eyes around and to aim the center of the eyes toward relevant parts of the environment.[64] This form of mapping, to the first approximation, can be described by the logarithmic transformation

$$x = l + m_x \ln r,$$
$$y = m_y \left(\theta + \frac{\pi}{2} \right). \tag{22}$$

Here, x and y are the cortical coordinates, r and θ are the polar coordinates in the retina, m_x and m_y are coefficients known as magnification factors, and l is a constant offset. Because the two halves of the two eyes converge in the same area of visual cor-

tex (Fig. 7A), the retinal polar angle varies between $-\pi/2$ and $\pi/2$.

It is interesting that the magnification factors along the two cortical directions introduced in Eq. 22 are found to be close to each other:[63]

$$m_x = m_y \equiv m. \tag{23}$$

This means that mapping between the eye and cortex is conformal and can be described by the complex logarithm transformation $x + iy = m \cdot \text{Log}\,(R_y + i R_x) + l$, where R_x and R_y are retinal coordinates. The Jacobian of this transformation, also known as the areal magnification factor, diverges near $r \equiv \sqrt{R_x^2 + R_y^2} = 0$, which means that the center of the eye is substantially overrepresented in the cortex compared to the periphery.

To implement this mapping, because the full distribution of molecular labels in the eye is not known, we will hypothesize one possible set of gradients that is consistent with existing data:[65]

$$q_A(r, \theta) = k_A e^{-r \cdot \cos \theta / d} + \varepsilon,$$
$$q_B(r, \theta) = k_B \left(\frac{\pi}{2} - \theta \right). \tag{24}$$

Here, k_A, k_B, and ε are some constants. The distribution of EphA receptor in the eye, therefore, decays exponentially away from the middle of the eye as a function of horizontal retinal coordinate $R_x = r \cdot \cos \theta$. This is consistent with existing observations.[65] The distribution of EphB molecules is assumed to be axial, that is, has a gradient oriented concentrically. As we explained earlier, two systems of gradients compete for the orientation of stripes, each preferring a stripe orientation parallel to it. Naturally, EphA receptors tend to orient stripes horizontally, while EphB gradients favor axial (concentric) stripe orientation. In the following, we will argue that EphA receptors win in the middle of the eye, while EphB receptors outcompete at the periphery, leading to the transition between horizontal and axial stripes somewhere in between.

To complete our argument, we need to postulate the distribution of ligands in the cortex. Because this distribution has not been measured quantitatively yet, we will adopt an approximation that is called the marker-induction model,[66] according to which the distribution of ligands is related to the profile of receptors through $\varphi_A \propto 1/q_A$ and $\varphi_B \propto q_B$. This approximation works to a degree to describe the distribution of molecules in lower vertebrates.[66] We will use

$$\varphi_A(x, y) = C_A/q_A(x, y),$$
$$\varphi_B(x, y) = \frac{C_B m}{k_B} q_B(x, y). \tag{25}$$

Here, C_A and C_B are the parameters of the marker-induction model. Having defined the gradients of molecular labels and the Hamiltonian, we can now calculate the orientation of thet stripes.

Orientation of ODPs is determined by finding the local $\vec{n}$ that minimizes ΔH and, consequently, $G(\vec{n})$ (Eq. 21). If $\vec{e}_x$ and $\vec{e}_y$ are unit vectors of x and y directions, the vector $\vec{n}$ can be written as $\vec{n} = \vec{e}_x \cos \Omega + \vec{e}_y \sin \Omega$.

$$\Omega_{\min} = -\frac{1}{2} \tan^{-1} \left[\frac{X \cdot \sin \frac{2y}{m}}{X \cdot \cos \frac{2y}{m} + Y} \right] + n\pi. \tag{26}$$

Here, $n = 0$ or $1/2$ is chosen to achieve minimum of ΔH. The auxiliary functions used in this expression are defined as follows:

$$X = \alpha C_A \left[e^{(x-l)/m} \frac{1}{dm} \frac{k_A}{k_A + \varepsilon e^{r \cdot \cos \theta/d}} \right]^2 \tag{27}$$

and

$$Y = \beta \frac{k_B C_B}{m}. \tag{28}$$

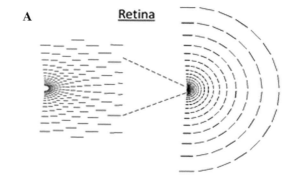

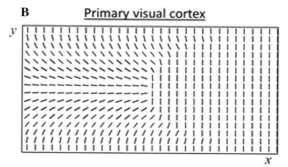

Figure 8. Numerical simulation of ODP orientations by minimizing energy function. Here, the parameters are $\alpha = \beta = 10$, $k_A = 1$, $k_B = 0.1$, $C_A = 10$, $C_B = 0.1$, $d = 4$, $l = 0$, and $\varepsilon = 0.1$. (A) Ocular dominance stripe orientations in retina generated by Eq. 26. (B) Stripe orientations in the cortex.

In Figure 8, we show the orientations of OD stripes in both cortex and retina calculated using Eq. 26. We find that these orientations capture prominent features of ODP orientation outlined in Figure 7(B). In particular, the orientation of OD stripes undergoes a transition from horizontal to concentric (Fig. 8). The point of transition can also be found to be

$$x_t = m_x \ln \left[-d \cdot W \left(-\frac{1}{\varepsilon} \sqrt{\frac{C_A \alpha}{m_x k_B C_B \beta}} \right) \right] + l. \tag{29}$$

Here, $W(x)$ is the Lambert function that can be found from $x = W e^W$. Overall, we showed that the model based on gradients of molecular labels can yield complex features of cortical maps, such as simultaneously formed maps of topography and OD. This argues that molecular gradients permit the formation of complex patterns of connections invoking a simple algorithm.

Simple mechanisms can explain the formation of complex features of connectivity

Here we argued that a 1-h instructional video could contain enough information to set up connections between billions of cortical neurons during neural development (see discussion after Eq. 4). To accomplish this task, the genome uses a strategy that relies on encoding general rules for wiring the neural net rather than specifying individual connections. One of the general rules uses the gradients of chemical labels to specify ordered connections for millions of axons in the visual system. We proposed that such a genetically specified mechanism can be combined with activity-dependent mechanisms that could potentially represent an organism's experience. We showed an example of such a unifying theory that uses a principle of the minimization of a cost function dependent on both chemical gradients and neural activity. As a corollary, our theory produced ODPs, a pattern of projections from two eyes into the same area. ODPs in our approach emerge from the interplay between the effects of molecular gradients and correlated neural activity. Thus, the patterns of OD observed in three-eyed frogs—surgically altered animals that are evolutionary naive to the problem of mixing projections from two eyes—could be viewed as a by-product of a mechanism that is based on both molecular gradients and neural activity. We showed that, within our model, the pattern of orientation of OD stripes observed in primates could be instructed to molecular gradients. Thus, the distributions of molecular cues could play instructive role for some features of ODPs.

The involvement of molecular labels in making connections between neurons has been proposed by Roger Sperry within the chemoaffinity hypothesis.[67] This hypothesis has withstood the test of time, with the discovery of many molecular cues, such as the Ephs and ephrins described here. The effects of correlated neural activity on the strengths of connections between neurons are described by Hebbian learning rules.[68] Here, we presented an approach that unifies Sperry's and Hebb's rules within a single mathematical model. Our model is based on the minimization of a cost function that we called a Hamiltonian. This approach to unifying disparate influences onto neural connections has been pio-

neered by Fraser and Perkel.[60] The details of the Hamiltonian are quite different in our theory from the Fraser and Perkel theory. For example, although Fraser and Perkel did report the emergence of ODPs in their Hamiltonian, ODPs are produced because of negative Fourier components of pair-wise interactions between axons, given by an oscillating Bessel function, that is, due to a mechanism similar to Turing instability.[53–56,59] The periodicity of ODPs in the previous approaches is therefore fixed by the range of interactions. In our theory, ODPs are formed by the interplay between featureless (Gaussian) interactions and gradients of molecular labels. The spacing of ODP in our model is not compatible with the spatial scales of either of these factors. Therefore, by changing the relative strengths of these factors, in our model, ODP periodicity can vary widely, as observed experimentally.[51] Nonetheless, our theory follows the spirit of the mathematical model proposed by Fraser and Perkel.

Although our theory is capable of explaining some experimental data, it certainly has limitations, mostly due to the simplifying assumptions that we had to make. For instance, we assumed that activity in the retina yields isotropic correlations (Eq. 10). If this assumption is relaxed, OD stripes may acquire preferred orientation. The effects of anisotropic correlations need to be further analyzed. Other gradients exist in the retina and in retina recipient regions, most notably ephrin ligands in retina and Eph receptors in retinal targets.[69] Gradients of other types of molecules may also play a role, such as Wnt, Ryk, and an engrailed protein En.[70,71] Finally, we hypothesized the 2D distribution of molecular labels in the primate visual system (Eq. 24), which should be experimentally testable.

Acknowledgments

This work was supported by NIH (NEI) R01EY018068, the Aspen Center for Physics, and the Swartz Foundation.

Conflict of interest

The authors declare no conflicts of interest.

References

1. Nicholls, J.G. 2012. *From Neuron to Brain.* Sunderland, MA: Sinauer Associates.
2. Kandel, E.R. 2012. *Principles of Neural Science.* New York: McGraw-Hill.

3. Squire, L.R. 2013. *Fundamental Neuroscience.* Amsterdam and Boston, MA: Elsevier/Academic Press.

4. Martin, S.J. P.D. Grimwood & R.G. Morris. 2000. Synaptic plasticity and memory: an evaluation of the hypothesis. *Annu. Rev. Neurosci.* **23:** 649–711.

5. Stevens, C.F. 1998. A million dollar question: does LTP = memory? *Neuron.* **20:** 1–2.

6. Cooke, S.F. & T.V. Bliss. 2006. Plasticity in the human central nervous system. *Brain: J. Neurol.* **129:** 1659–1673.

7. Fusi, S. P.J. Drew & L.F. Abbott. 2005. Cascade models of synaptically stored memories. *Neuron.* **45:** 599–611.

8. Lisman, J.E. & A.M. Zhabotinsky. 2001. A model of synaptic memory: a CaMKII/PP1 switch that potentiates transmission by organizing an AMPA receptor anchoring assembly. *Neuron.* **31:** 191–201.

9. Miller, P. *et al.* 2005. The stability of a stochastic CaMKII switch: dependence on the number of enzyme molecules and protein turnover. *PLoS Biol.* **3:** e107.

10. Petersen, C.C. *et al.* 1998. All-or-none potentiation at CA3–CA1 synapses. *Proc. Natl. Acad. Sci. U. S. A.* **95:** 4732–4737.

11. Enoki, R. *et al.* 2009. Expression of long-term plasticity at individual synapses in hippocampus is graded, bidirectional, and mainly presynaptic: optical quantal analysis. *Neuron.* **62:** 242–253.

12. Escobar, G. T. Fares & A. Stepanyants. 2008. Structural plasticity of circuits in cortical neuropil. *J. Neurosci.: Off. J. Soc. Neurosci.* **28:** 8477–8488.

13. Bliss, T.V. & G.L. Collingridge. 2013. Expression of NMDA receptor-dependent LTP in the hippocampus: bridging the divide. *Mol. Brain.* **6:** 5.

14. Braitenberg, V. & A. Schüz. 1998. *Cortex: Statistics and Geometry of Neuronal Connectivity.* Berlin and New York: Springer.

15. Wen, Q. & D.B. Chklovskii. 2005. Segregation of the brain into gray and white matter: a design minimizing conduction delays. *PLoS Computat. Biol.* **1:** e78.

16. Rodieck, R.W. 1998. *The First Steps in Seeing.* Sunderland, MA: Sinauer Associates.

17. Knott, G.W. *et al.* 2006. Spine growth precedes synapse formation in the adult neocortex in vivo. *Nature Neurosci.* **9:** 1117–1124.

18. Trachtenberg, J.T. *et al.* 2002. Long-term in vivo imaging of experience-dependent synaptic plasticity in adult cortex. *Nature.* **420:** 788–794.

19. Alvarez, V.A. & B.L. Sabatini. 2007. Anatomical and physiological plasticity of dendritic spines. *Annu. Rev. Neurosci.* **30:** 79–97.

20. Fu, M. & Y. Zuo. 2011. Experience-dependent structural plasticity in the cortex. *Trends Neurosci.* **34:** 177–187.

21. Grutzendler, J. N. Kasthuri & W.B. Gan. 2002. Long-term dendritic spine stability in the adult cortex. *Nature.* **420:** 812–816.

22. Feldman, D.E. 2009. Synaptic mechanisms for plasticity in neocortex. *Annu. Rev. Neurosci.* **32:** 33–55.

23. Cover, T.M. & J.A. Thomas. 2006. *Elements of Information Theory.* Hoboken, NJ: Wiley-Interscience.

24. Kolterman, B.E. & A.A. Koulakov. 2010. Is universal coverage good for neurons? *Neuron.* **66:** 1–3.

25. Chklovskii, D.B. & A.A. Koulakov. 2004. Maps in the brain: what can we learn from them? *Annu. Rev. Neurosci.* **27:** 369–392.

26. McLaughlin, T. & D.D. O'Leary. 2005. Molecular gradients and development of retinotopic maps. *Annual review of neuroscience.* **28:** 327–355.

27. Triplett, J.W. *et al.* 2009. Retinal input instructs alignment of visual topographic maps. *Cell.* **139:** 175–185.

28. Kalatsky, V.A. & M.P. Stryker. 2003. New paradigm for optical imaging: temporally encoded maps of intrinsic signal. *Neuron.* **38:** 529–545.

29. Flanagan, J.G. & P. Vanderhaeghen. 1998. The ephrins and Eph receptors in neural development. *Annu. Rev. Neurosci.* **21:** 309–345.

30. O'Leary, D.D. & T. McLaughlin. 2005. Mechanisms of retinotopic map development: Ephs, ephrins, and spontaneous correlated retinal activity. *Prog. Brain Res.* **147:** 43–65.

31. O'Leary, D.D.M. & D.G. Wilkinson. 1999. Eph receptors and ephrins in neural development. *Curr. Op. Neurobiol.* **9:** 65–73.

32. McLaughlin, T. *et al.* 2003. Bifunctional action of ephrin-B1 as a repellent and attractant to control bidirectional branch extension in dorsal-ventral retinotopic mapping. *Development.* **130:** 2407–2418.

33. Tsigankov, D.N. & A.A. Koulakov. 2004. Can repulsion be induced by attraction: a role of ephrin-B1 in retinotectal mapping? *Arxiv q-bio:* xxx.lanl.gov/abs/q-bio/0403013.

34. Triplett, J.W. *et al.* 2011. Competition is a driving force in topographic mapping. *Proc. Natl. Acad. Sci. U. S. A.* **108:** 19060–19065.

35. Koulakov, A.A. & D.N. Tsigankov. 2004. A stochastic model for retinocollicular map development. *BMC Neurosci.* **5:** 30.

36. Tsigankov, D.N. & A.A. Koulakov. 2006. A unifying model for activity-dependent and activity-independent mechanisms predicts complete structure of topographic maps in ephrin-A deficient mice. *J. Computat. Neurosci.* **21:** 101–114.

37. Tsigankov, D. & A. Koulakov. 2009. Optimal axonal and dendritic branching strategies during the development of neural circuitry. *Front. Neural Circ.* **3:** 18.

38. Tsigankov, D. & A.A. Koulakov. 2010. Sperry versus Hebb: topographic mapping in Isl2/EphA3 mutant mice. *BMC Neurosci.* **11:** 155.

39. Cang, J. *et al.* 2005. Ephrin-as guide the formation of functional maps in the visual cortex. *Neuron.* **48:** 577–589.

40. McLaughlin, T. *et al.* 2003. Retinotopic map refinement requires spontaneous retinal waves during a brief critical period of development. *Neuron.* **40:** 1147–1160.

41. Cang, J. *et al.* 2005. Development of precise maps in visual cortex requires patterned spontaneous activity in the retina. *Neuron.* **48:** 797–809.

42. Ackman, J.B. T.J. Burbridge & M.C. Crair. 2012. Retinal waves coordinate patterned activity throughout the developing visual system. *Nature.* **490:** 219–225.

43. Demas, J.A. H. Payne & H.T. Cline. 2012. Vision drives correlated activity without patterned spontaneous activity in developing Xenopus retina. *Dev. Neurobiol.* **72:** 537–546.

44. Xu, H.P. *et al.* 2011. An instructive role for patterned spontaneous retinal activity in mouse visual map development. *Neuron.* **70:** 1115–1127.

45. Snider, J. A. Pillai & C.F. Stevens. 2010. A universal property of axonal and dendritic arbors. *Neuron.* **66:** 45–56.

46. Shatz, C.J. & M.P. Stryker. 1978. Ocular dominance in layer IV of the cat's visual cortex and the effects of monocular deprivation. *J. Physiol.* **281:** 267–283.

47. Horton, J.C. & D.R. Hocking. 1996. Intrinsic variability of ocular dominance column periodicity in normal macaque monkeys. *J. Neurosci.: Off. J. Soc. Neurosci.* **16:** 7228–7239.

48. LeVay, S. M.P. Stryker & C.J. Shatz. 1978. Ocular dominance columns and their development in layer IV of the cat's visual cortex: a quantitative study. *J. Comp. Neurol.* **179:** 223–244.

49. Chklovskii, D.B. & A.A. Koulakov. 2000. A wire length minimization approach to ocular dominance patterns in mammalian visual cortex. *Physica A.* **284:** 318–334.

50. Koulakov, A.A. & D.B. Chklovskii. 2003. Ocular dominance patterns and the wire length minimization: a numerical study. *Arxiv q-bio:* xxx.lanl.gov/abs/q-bio/0311027.

51. Hensch, T.K. & M.P. Stryker. 2004. Columnar architecture sculpted by GABA circuits in developing cat visual cortex. *Science.* **303:** 1678–1681.

52. Chklovskii, D.B. 2000. Binocular disparity can explain the orientation of ocular dominance stripes in primate primary visual area (V1). *Vis. Res.* **40:** 1765–1773.

53. Miller, K.D. J.B. Keller & M.P. Stryker. 1989. Ocular dominance column development: analysis and simulation. *Science.* **245:** 605–615.

54. Swindale, N.V. 1980. A model for the formation of ocular dominance stripes. *Proc. R. Soc. Lond. B. Biol. Sci.* **208:** 243–264.

55. Swindale, N.V. 1996. The development of topography in the visual cortex: a review of models. *Network.* **7:** 161–247.

56. Willshaw, D.J. & C. von der Malsburg. 1976. How patterned neural connections can be set up by self-organization. *Proc. R. Soc. Lond. B. Sci.* **194:** 431–445.

57. Cline, H.T. E.A. Debski & M. Constantine-Paton. 1987. N-methyl-D-aspartate receptor antagonist desegregates eye-specific stripes. *Proc. Natl. Acad. Sci. U. S. A.* **84:** 4342–4345.

58. Reh, T.A. & M. Constantine-Paton. 1985. Eye-specific segregation requires neural activity in three-eyed Rana pipiens. *J. Neurosci.: Off. J. Soc. Neurosci.* **5:** 1132–1143.

59. Cross, M.C. & P.C. Hohenberg. 1993. Pattern-formation outside of equilibrium. *Rev. Mod. Phys.* **65:** 851–1112.

60. Fraser, S.E. & D.H. Perkel. 1990. Competitive and positional cues in the patterning of nerve connections. *J. Neurobiol.* **21:** 51–72.

61. Kanold, P.O. *et al.* 2009. Co-regulation of ocular dominance plasticity and NMDA receptor subunit expression in glutamic acid decarboxylase-65 knock-out mice. *J. Physiol.* **587:** 2857–2867.

62. Turrigiano, G.G. 1999. Homeostatic plasticity in neuronal networks: the more things change, the more they stay the same. *Trends Neurosci.* **22:** 221–227.

63. LeVay, S. *et al.* 1985. The complete pattern of ocular dominance stripes in the striate cortex and visual field of the macaque monkey. *J. Neurosci.: Off. J. Soc. Neurosci.* **5:** 486–501.

64. Koulakov, A.A. 2010. On the scaling law for cortical magnification factor. *Arxiv q-bio:* xxx.lanl.gov/abs/1002.4368.

65. Lambot, M.A. *et al.* 2005. Mapping labels in the human developing visual system and the evolution of binocular vision. *J. Neurosc.: Off. J. Soc. Neurosci.* **25:** 7232–7237.

66. Willshaw, D. 2006. Analysis of mouse EphA knockins and knockouts suggests that retinal axons programme target cells to form ordered retinotopic maps. *Development.* **133:** 2705–2717.

67. Sperry, R.W. 1963. Chemoaffinity in the orderly growth of nerve fiber patterns and connections. *Proc. Natl. Acad. Sci. U.S.A.* **50:** 703–710.

68. Hebb, D.O. 1949. *The Organization of Behavior: A Neuropsychological Theory.* New York: Wiley.

69. Rashid, T. *et al.* 2005. Opposing gradients of ephrin-as and epha7 in the superior colliculus are essential for topographic mapping in the mammalian visual system. *Neuron.* **47:** 57–69.

70. Schmitt, A.M. *et al.* 2006. Wnt-Ryk signalling mediates medial-lateral retinotectal topographic mapping. *Nature.* **439:** 31–37.

71. Brunet, I. *et al.* 2005. The transcription factor Engrailed-2 guides retinal axons. *Nature.* **438:** 94–98.

Ann. N.Y. Acad. Sci. ISSN 0077-8923

ANNALS OF THE NEW YORK ACADEMY OF SCIENCES
Issue: *Cracking the Neural Code: Third Annual Aspen Brain Forum*

In vivo robotics: the automation of neuroscience and other intact-system biological fields

Suhasa B. Kodandaramaiah,[1,2] Edward S. Boyden,[2] and Craig R. Forest[1]

[1]George W. Woodruff School of Mechanical Engineering, Georgia Institute of Technology, Atlanta, Georgia. [2]MIT Media Lab and McGovern Institute, Department of Biological Engineering and Department of Brain and Cognitive Sciences, Massachusetts Institute of Technology, Cambridge, Massachusetts

Address for correspondence: Craig R. Forest, George W. Woodruff School of Mechanical Engineering, Georgia Institute of Technology, 813 Ferst Dr., Room 411, Atlanta, GA 30332. cforest@gatech.edu; and Edward S. Boyden, MIT Media Lab and McGovern Institute, MIT, E15-421, 20 Ames St., Cambridge, MA 02139. esb@media.mit.edu

Robotic and automation technologies have played a huge role in *in vitro* biological science, having proved critical for scientific endeavors such as genome sequencing and high-throughput screening. Robotic and automation strategies are beginning to play a greater role in *in vivo* and *in situ* sciences, especially when it comes to the difficult *in vivo* experiments required for understanding the neural mechanisms of behavior and disease. In this perspective, we discuss the prospects for robotics and automation to influence neuroscientific and intact-system biology fields. We discuss how robotic innovations might be created to open up new frontiers in basic and applied neuroscience and present a concrete example with our recent automation of *in vivo* whole-cell patch clamp electrophysiology of neurons in the living mouse brain.

Keywords: robotics; neuroscience; patch clamping

Introduction: automation in biology

Robotics has played a major role in the advancement of biological research in the past few decades. Semi-autonomous machines integrate hardware, wetware, and software from precision engineered or microfabricated parts to nimbly load, manipulate, and measure thousands to millions of biological samples simultaneously, more rapidly, more sensitively, more accurately, or in a more repeatable manner than manual approaches. Their applications span the research space from automated phenotyping to high-throughput screening to imaging to genome sequencing. Examples abound for how these tools have opened the door to comprehensive biological studies. In the race to sequence the human genome in the 1990s, robots capable of high-throughput analysis of microliter volumes of liquid were developed and deployed massively (Fig. 1). As just a few examples, high-throughput microfluidics[1,2] can now be used to perform nearly 10,000 independent real-time polymerase chain reactions (PCR) for genotyping[3] and transcriptome profiling applications.[4–8] Fluid-handling robots have revolutionized synthetic biology by enabling the efficient, rapid transfer of reagents from one set of plates to another.[9,10] Automated plate readers and microscopy methods enable time-lapse imaging of physiological changes in cultured cells, and *in vitro* patch clamping enables automated electrophysiology in cell lines.[11–15] These innovations are very widely used, and sometimes ubiquitous, in major research institutions and industrial settings such as in pharmaceutical and biotechnology companies.

Value and principles of *in vivo* and *in situ* automation

Despite this progress and the pervasive presence of automation in molecular biology today, there remain many tedious and repetitive manual tasks, as well as more complex tasks that defy straightforward automation, and that are more akin to art forms than scientific processes. Often they have not been systematically analyzed but are passed down through generations of researchers as best practices. This practice sometimes limits use to a few highly skilled laboratories, especially when living organs or

doi: 10.1111/nyas.12171

Ann. N.Y. Acad. Sci. 1305 (2013) 63–71 © 2013 New York Academy of Sciences.

Figure 1. One of many rows of ABI 3730xl automated DNA Analyzers for shotgun sequencing of the human genome in months (30 billions bp/year) in 2005. (Courtesy: Steve Jurvetson.)

organisms are involved, as in *in vivo* neuroscience experiments. This frontier is an opportunity because automation not only makes things simpler, increasing the rate of adoption and progress of specific approaches, but also broadens the number of individuals and laboratories that can contribute innovations, since they no longer have to improve an art form, but rather can iteratively improve the protocols performed by a robotic agent. Further, higher throughput automation can enable scaling of the number of parallel samples being analyzed, or the number of analyses being performed per sample, or the sampling rate. It can also lead to increased standardization of procedures across laboratories, important for improving the ability of different parts of the literature to be integrated, and for results from different groups to be compared.

In many cases, increasing the number of observations that can be made in parallel is important not only for augmenting the amount of data that can be collected, but also for opening up fundamentally new kinds of investigation. For example, making

many simultaneous observations on different parts of an intact system—an organ, such as the brain, or even an entire organism—might reveal correlated, and perhaps coordinated, physiological processes taking place in different parts of the system, which would never be revealed by investigation of single sites one at a time. As another example, the ability to perform tasks often done *in vitro*—from pharmacological assessment to biochemical analysis—in the living organism, would enable detailed understanding of how specific processes of basic or applied scientific interest take place, in the full context of an intact organism, including its baseline activity, awake or behaving state–dependent modulation, or disease states.

Rethinking *in vivo* procedures for automation

In order to automate a complex *in vivo* or *in situ* procedure, it is very important to understand how humans perform the procedure, not only analyzing the procedure at face value but also delving deeply into the parameters that govern success or failure of the procedure. This often means that the engineer seeking to automate a procedure must not simply take requests from biologists and merely attempt to automate their stated methods, but instead must herself master the procedure, so that it is possible for the engineer to understand the best possible way for the procedure to be automated. Very often, the way that a human performs will not necessarily be the easiest way to perform the task in an automated fashion. Humans often use complex sets of cues— visual and auditory, for example—to perform *in vivo* experiments. But for automation, it may be important to rely upon more straightforward and less complex modalities, such as electrical impedance (which can nevertheless indicate important properties of a tissue that a robotic device is exploring). In addition, humans conduct *in vivo* experiments with their very-high-degree-of-freedom hands and synergistic muscular systems, but the range of inexpensive actuators that are reliable and inexpensive enough to become commonly used in messy and complex biology lab environments, may require a greater reliance on good software and rethinking of the procedure to minimize the number of expensive actuators required, or even a rethinking of the modality of actuation (e.g., replacing a complex robot arm with a scanning laser beam). Later, we explore these two arenas of endeavor—how to find the most easily automated methodology for performing a complex *in vivo* procedure, and how to devise the simplest and most robust modality of actuation and style of robot—in the context of an area that we have recently pioneered, the automation of intracellular neural recording in the living brain.

In vivo neuroscience and past automation efforts

The vertebrate brain is a complex organ consisting of billions of neurons,[16] each of which is interconnected with thousands of other neurons through synapses.[17] Each neuron receives information via synaptic transmission, computes an electrical signal within it, and transmits information to downstream neurons. They express different sets of genes,[18] have myriad morphologies, and undergo plasticity in different ways during performance of cognitive tasks and learning. Thus, one of the fundamental challenges for neuroscientists has been the difficulty of linking the knowledge we have on cellular-level

phenomena, such as synaptic transmission, often gained by manual *in vitro* experimental preparations; to emergent properties of the intact living system such as learning and memory. Technologies including electrical neural recording, the generation of transgenic animals, the use of optogenetic neural control,[19,20] optics to image intact neural systems, and cell- and circuit-resolution molecular and biochemical analyses are all important. However, many of these techniques are art forms, requiring extensive effort to learn, typically time consuming to perform, and lacking in scale, without standardization across groups, and with innovations typically driven in different directions by different laboratories. Arguably, many of these areas are ripe for robotic innovation.

In these *in vivo* and *in situ* (i.e., intact tissue) spaces of endeavor, robotics and automation have already begun to make inroads. The use of laser capture microdissection to automatically isolate cellular contents from tissues is enabling new kinds of systems biology,[21] and also supporting a diversity of histopathological studies. Microfluidic devices for whole-organism imaging and sorting are having great impact on the study of organisms, such as *Caenorhabditis elegans*,[22–25] Drosophila,[26] and zebrafish,[27] enabling rapid imaging, sorting, and adaptive control of these organisms for both advancement of basic biology as well as accelerated pharmacological screening. Another area where robotics has played a crucial role for *in situ* analysis is in the field of intact tissue imaging. Automated serial block-face scanning electron microscopy, as well as automated histology systems,[28] have driven progress in the nascent field of connectomics.[29] High-throughput, automated *in situ* hybridization along with automated imaging platforms were indispensable for charting the mouse brain gene expression maps of the Allen Brain Atlas.[30] Intact tissue analysis has benefited from automated sectioning, *in situ* hybridization, and imaging of tissue samples.[28,31–33] Motorized devices have been devised to support the lowering of tetrodes into the living rat brain,[34] and to enable automated stabilization of extracellular recording electrodes for maintaining optimal recording quality.[35] In addition, automated electrode recording stabilization techniques have been explored for stabilizing sharp recordings against brain movement in awake, behaving zebra finches.[36] We have recently explored the automation of multisite viral injection, using precisely timed fluidic delivery of viruses to three-dimensional structures in the brain, in an easily user-customizable fashion[37] (a process that one can imagine would easily be extended to stem cell or pharmacological injection in many sites at once). Microfabricated strategies for adaptively moving many extracellular microelectrodes could lead to improvements in the development of reliable and stable interfaces with single neurons, important for basic neurophysiological studies and emerging cortical prosthetic technologies.[38] Beyond the analysis of live animals and preserved animal and human tissues, robotic actuators are now routinely used in clinical settings to enhance the ability of humans to perform complex surgical procedures.[39] For example, the Amadeus and Da Vinci robotics are used to perform minimally invasive laproscopic procedures, as well as robot-assisted telesurgery.[40] Robots are increasingly playing a role in even delicate neurosurgeries, as well as in cardiac surgery,[41] and are being incorporated into operating room systems that enable noninvasive visualization (e.g., the neuroArm, an MRI-guided robotic actuator[42]). These advances illustrate the broad and deep impacts already stemming from automation technologies on the study or manipulation of living or intact biological systems in basic biology and medicine.

Case study: *in vivo* patch clamp neural recording

To explore in depth a specific avenue of *in vivo* robotic engineering, we discuss a technology that we recently developed that assists in the mechanistic understanding of how cellular-level activities of neuronal networks give rise to higher level cognitive abilities, and how they go awry in brain disorders. To study cellular-level activity, ideally one would be able to observe electrical activities in neurons with intracellular, synaptic resolution, and ideally in a fashion capable of linking this physiological information to the genetic and morphological information associated with the cellular identity. Such integrated network-wide studies will require new technologies that can access these single cells efficiently and in way that is able to be scaled. One method for doing this, which works even *in vivo*, is whole-cell patch clamp neural recording. In this technique, a glass micropipette establishes an electrical and molecular connection to the insides of

an individual cell embedded in intact tissue. Invented in 1981,[43] and winning Neher and Sakmann the Nobel Prize in 1991, whole-cell patch clamping enables recording of the electrical activity of neurons *in vivo* that exhibit signal quality and temporal fidelity sufficient to report synaptic and ion channel–mediated subthreshold events of importance for understanding not only how neurons compute during behavior, but also how their physiology changes in disease states or in response to drug administration. Further, it enables dye infusion for morphological visualization, and extraction of cell contents for transcriptomic analysis.[44–46] Potentially, *in vivo* patch clamping could have clinical impact, being used in neurosurgical settings to do integrative measurements on single cells in, for example, the epileptic brain. This technique could also be used to analyze biopsy samples (e.g., from tumors) or from surgical resections of the brain, at single-cell resolution, potentially revealing tissue heterogeneity and thus new principles of personalized medicine. Such studies that link molecular, cellular, and anatomical properties of individual cells to their behavioral or disease circuit context are difficult *in vivo*. Despite these compelling opportunities, *in vivo* patching requires skill, and the hardware required is specialized and expensive. Thus, *in vivo* patching has been utilized by a relatively small number of labs, and is usually regarded as a difficult technique, performed manually by highly skilled operators trained by masters, in the anesthetized brain, and in very limited applications in the awake brain.[47–52] Accordingly, we considered it as an ideal example of a neuroscience technique to automate.

We began by analyzing what humans do while they perform *in vivo* patch clamping. This required examination of humans in the laboratory as they performed patch clamping, to analyze the actual methodology they used to perform this task. Importantly, we then examined the physics and mechanics of what was being done, in order to isolate parameters most amenable to automation (e.g., focusing on a time-series electrical impedance analysis for our automation, rather than relying on visual detection of stereotyped patterns of electrical signal, such as heartbeat-rhythm modulation of the recording).

By doing this analysis, we discovered that single cells could be accurately detected by analyzing the temporal sequence of micropipette impedance

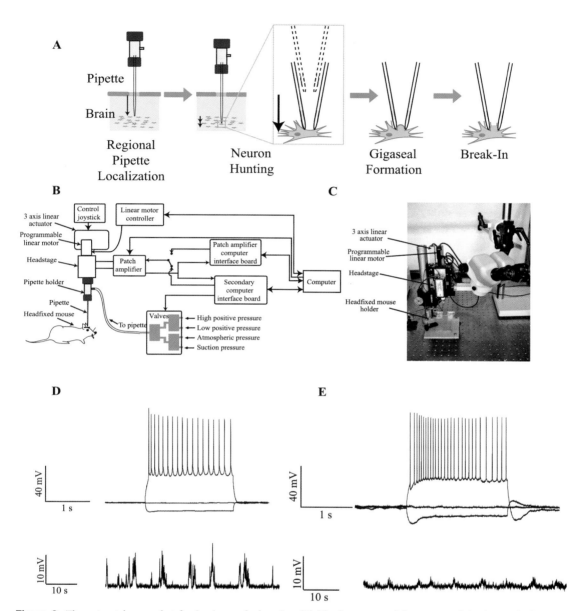

Figure 2. The autopatcher: a robot for *in vivo* patch clamping. (A) The four stages of the automated *in vivo* patch algorithm, discovered through iterative exploration of the parameters governing successful patch clamping: regional pipette localization, in which the pipette is lowered to a target zone in the brain; neuron hunting, in which the pipette is advanced until a neuron is detected via a change in pipette resistance; gigaseal formation, in which a gigaseal state is achieved (if cell-attached patching is desired, the algorithm can end here); break-in, in which the whole-cell state is achieved. (B) Schematic of a simple robotic system capable of performing the autopatching algorithm. The system consists of a conventional *in vivo* patch setup (i.e., pipette, headstage, three-axis linear actuator, patch amplifier plus computer interface board, and computer), equipped with a few additional modules: a programmable linear motor (to move the pipette up and down in a temporally precise fashion), a controllable bank of pneumatic valves for pressure control, and a secondary computer interface board to enable closed-loop control of the motor based upon sequences of pipette resistance measurements. (C) Photograph of the setup, focusing on three-axis linear actuator (with additional programmable linear motor) and the holder for head fixing the mouse. (D) Current clamp traces during current injection for a cortical neuron for which whole-cell state was established via the autopatcher. (E) Current clamp traces during current injection for a hippocampal neuron for which whole-cell state was established via the autopatcher. Adapted, with permission, from Kodandaramaiah *et al.*[53]

changes as the micropipette is lowered into the brain (Fig. 2A), looking for particular signatures of temporal change in pipette resistance. Building from this observation, we found that blind *in vivo* whole-cell patching of neurons, in which micropipettes were lowered until a cell is detected and then recorded, could be reduced to a reliable algorithm, in which cells are detected with >90% yield, and the whole-cell state established in 40–60% of detected cells,[53] with the yields for whole-cell state establishment exceeding 60–70% at the beginning of a session when the brain was intact, and declining to 30–50% over time as multiple brain penetrations occurred.

The next step was to devise the simplest robotic invention that could perform this algorithm. The algorithm could be realized by a simple robot (Fig. 2B and C), which actuates a set of motors and valves rapidly upon recognition of specific temporal sequences of micropipette impedance changes, achieving whole-cell patch clamp recordings in 3–7 min each in the live mouse brain. The robot we designed is relatively inexpensive, is made out of easily accessible, commercially available parts, and thus can easily be appended to an existing patch clamp electrophysiology rig, by adding a few valves, a computer-controlled linear motor, and a digital interface board for analyzing micropipette impedance changes, and actuating the motor and valves.

The robot can obtain very high quality intracellular electrical recordings of neurons, even millimeters deep, in living mouse brain (Fig. 2E), and works in multiple brain regions, suggesting that our algorithm has a degree of generality. The fact that a robot performs it makes exploration of new algorithm variants a simple task. We are starting to explore the capabilities of our robot in the awake rodent brain[54] and its compatibility with optogenetics, which could enable on-the-fly cell type identification.[54] We are currently aiming to derive new algorithms for procedures that humans do not perform at all, such as the simultaneous patch clamp of many neurons at once,[55,56] which requires study of how multiple independent patch pipettes might interact in the dense tissue of the living brain.

Potential pitfalls in *in vivo* robotics and how they might be addressed

In vivo neuroscience is a combination of multiple experimental procedures, often required to be performed sequentially. For fully realizing the potential of robotics, it is not necessary just to automate one aspect of the experimental workflow, but develop platforms for automating upstream and downstream processes as well. These systems will need to be built modularly, so as to be easily integrated into the work flow and reconfigurable to be broadly applicable to many experimental protocols, while taking into consideration inherent variability in the *in vivo* biological milieu. For example, in the case study of our automated patch clamp robot, true high throughput will only be achieved if we can scale the control algorithms and the hardware to control large arrays of recording electrodes in the fashion of tetrodes and of additionally automated accessory tasks, such as animal surgery, the fabrication of electrodes, the swapping in of fresh electrodes after each experiment, and real-time logging and analyzing of the acquired data. In particular, as automation of other procedures such as animal neurosurgeries emerge, ethical considerations will be important in the engineering as well—for example, if the robot were to fail, it should fail in a safe mode that does not jeopardize the health or well-being of the animal subject. Also, if a robot must halt a procedure that is partially completed, it should be able to either bring the experimental episode to a conclusion in a way that preserves animal welfare, or promptly alert a nearby human attendant to intervene. One possibility is that *in vivo* automation will enable scientists to perform experiments much more efficiently and effectively than before, enabling higher success rates in animal experimentation and collection of more data per animal used, important for ethical, scientific, and financial reasons.

Endeavors in *in vivo* robotics may benefit from multilab collaborations that collectively contain deep domain knowledge in multiple aspects of neuroscience as well as multiple aspects of the engineering. These efforts will require significant investments for innovating technologies and scaling up infrastructure as technologies mature. Recent work by the Allen Institute for Brain Science,[30] the emerging European Human Brain Project,[57] and the excitement generated by recent proposals in the United States, such as the Brain Activity Map initiative,[58–60] suggest that the time may be ripe for the automation and scaling up of neuroscientific procedures.

This also brings into discussion the role that will have to be played by entities outside of academia. Although many entities have launched to foster the distribution of DNA (e.g., http://addgene.org) and viruses (e.g., various university viral core facilities), there are no comparable methodologies for distributing robots. Entrepreneurial and commercialization endeavors may be the *de facto* path, but open access and open source models may well be as important (if not more important, in the early days when many groups may seek to customize robots for specific kinds of applications). Ensuring broad dissemination means increasing usability (i.e., through the creation of simple yet powerful user interfaces, working to make devices fault tolerant, and connecting devices to existing laboratory hardware), while working to keep costs and prices down to maximize impact. To enable rapid dissemination to the scientific community, it may be important that commercial entities pursue such simplification and robustness activities, while simultaneously enabling immediate free access to nonprofit and academic researchers who seek to try it out right away (e.g., by making all-parts lists, computer-aided design drawings, and software available on the internet, or even by setting up core facilities within universities to manufacture components for their communities).

There may well be great demand in the future for innovation in the field of *in vivo* robotics, particularly in neuroscience. *In vivo* stem cell biology, *in vivo* imaging (especially over long periods of time), stereotactic surgery (to insert drugs and devices, or to make measurements), and *ex vivo* analyses of tissues are all in need of automation that powerfully enables everyday art forms to become simple and inexpensive, and that later enable these tasks to be performed at such scales, and in such integrated fashions, as to reveal new kinds of integrated patterns and principles of biological operation. As such *in vivo* automation tools find uses driven by biological discovery, it is likely that they, like *in vitro* automation tools, will find clinical uses, perhaps in contexts such as diagnostics, neurosurgery, or other fields.

Acknowledgments

E.S.B. and C.R.F. acknowledge NIH Single Cell Grant 1 R01 EY023173. E.S.B. acknowledges Human Frontiers Science Program; IET A. F. Harvey Prize; MIT McGovern Institute and McGovern Institute Neurotechnology (MINT) Program; MIT Media Lab and Media Lab Consortia; New York Stem Cell Foundation–Robertson Investigator Award; NIH Director's New Innovator Award 1DP2OD002002, NIH EUREKA Award 1R01NS075421, NIH Transformative R01 1R01GM104948, and NIH Grants 1R01DA029639, and 1R01NS067199; NSF CAREER Award CBET 1053233 and DMS1042134 (the Cognitive Rhythms Collaborative); Paul Allen Distinguished Investigator in Neuroscience Award; Shelly Razin; and SkTech. C.R.F. acknowledges funding from NSF (EHR 0965945 and CISE 1110947), NIH Computational Neuroscience Training grant (DA032466–02), Georgia Tech Translational Research Institute for Biomedical Engineering & Science (TRIBES) Seed Grant Awards Program, Wallace H. Coulter Translational/Clinical Research Grant Program, and support from Georgia Institute of Technology through the Institute for Bioengineering and Biosciences Junior Faculty Award, Technology Fee Fund, Invention Studio, and the George W. Woodruff School of Mechanical Engineering.

Conflicts of interest

C.R.F., E.S.B., and S.B.K. are coinventors on a patent owned by MIT and the Georgia Institute of Technology. C.R.F. and S.B.K. are financially affiliated with Neuromatic Devices, which is seeking to manufacture and sell autopatching robots.

References

1. Thorsen, T., S.J. Maerkl & S.R. Quake. 2002. Microfluidic large-scale integration. *Science* **298:** 580–584.
2. White, A.K. *et al.* 2011. High-throughput microfluidic single-cell RT-qPCR. *Proc. Natl. Acad. Sci. USA* **108:** 13999–14004.
3. Fan, H.C., J. Wang, A. Potanina & S.R. Quake. 2011. Whole-genome molecular haplotyping of single cells. *Nat. Biotech.* **29:** 51–57.
4. Sanchez-Freire, V., A.D. Ebert, T. Kalisky, *et al.* 2012. Microfluidic single-cell real-time PCR for comparative analysis of gene expression patterns. *Nat. Protocols* **7:** 829–838.
5. Warren, L., D. Bryder, I.L. Weissman & S.R. Quake. 2006. Transcription factor profiling in individual hematopoietic progenitors by digital RT-PCR. *Proc. Natl. Acad. Sci.* **103:** 17807–17812.
6. Dalerba, P. *et al.* 2011. Single-cell dissection of transcriptional heterogeneity in human colon tumors. *Nat. Biotech.* **29:** 1120–1127.
7. Citri, A., Z.P. Pang, T.C. Sudhof, *et al.* 2012. Comprehensive qPCR profiling of gene expression in single neuronal cells. *Nat. Protocols* **7:** 118–127.

8. Teh, S.Y., R. Lin, L.H. Hung & A.P. Lee. 2008. Droplet microfluidics. *Lab. Chip* **8:** 198–220.

9. Wang, H.H. *et al.* 2009. Programming cells by multiplex genome engineering and accelerated evolution. *Nature* **460:** 894–898.

10. Tian, J. 2009. DNA synthesis. In *Automation in Proteomics and Genomics: An Engineering Case-Based Approach.* G. Alterovitz, R. Benson & M. Ramoni, Eds.: 175–191. Chichester: John Wiley & Sons, Ltd.

11. Dunlop, J., M. Bowlby, R. Peri, *et al.* 2008. High-throughput electrophysiology: an emerging paradigm for ion-channel screening and physiology. *Nat. Rev. Drug Discov.* **7:** 358–368.

12. Estacion, M. *et al.* 2010. Can robots patch-clamp as well as humans? Characterization of a novel sodium channel mutation. *J. Physiol.* **588:** 1915–1927.

13. Milligan, C.J. *et al.* 2009. Robotic multiwell planar patch-clamp for native and primary mammalian cells. *Nat. Protocols* **4:** 244–255.

14. Wang, X. & M. Li. 2004. Automated electrophysiology: high throughput of art. *Assay Drug Dev. Technol.* **1:** 695–708.

15. Brueggemann, A., M. George, M. Klau, *et al.* 2004. Ion channel drug discovery and research: the automated Nano-Patch-Clamp technology. *Curr. Drug Discov. Technol.* **1:** 91–96.

16. Nowakowski, R.S. 2006. Stable neuron numbers from cradle to grave. *Proc. Natl. Acad. Sci.* **103:** 12219–12220.

17. Drachman, D.A. 2005. Do we have brain to spare? *Neurology* **64:** 2004–2005.

18. Markram, H. *et al.* 2004. Interneurons of the neocortical inhibitory system. *Nat. Rev. Neurosci.* **5:** 793–807.

19. Boyden, E.S. 2011. A history of optogenetics: the development of tools for controlling brain circuits with light. *F1000 Biol. Rep.* **3:** 11.

20. Bernstein, J.G., P.A. Garrity & E.S. Boyden. 2011. Optogenetics and thermogenetics: technologies for controlling the activity of targeted cells within intact neural circuits. *Curr. Opin. Neurobiol.* **22:** 61–71.

21. Boyden, E.S. *et al.* 2006. Selective engagement of plasticity mechanisms for motor memory storage. *Neuron* **51:** 823–834.

22. Kocabas, A., C.H. Shen, Z.V. Guo & S. Ramanathan. 2012. Controlling interneuron activity in Caenorhabditis elegans to evoke chemotactic behaviour. *Nature* **490:** 273–277.

23. Leifer, A.M., C. Fang-Yen, M. Gershow, *et al.* 2011. Optogenetic manipulation of neural activity in freely moving *Caenorhabditis elegans. Nat. Methods* **8:** 147–152.

24. Samara, C. *et al.* 2010. Large-scale *in vivo* femtosecond laser neurosurgery screen reveals small-molecule enhancer of regeneration. *Proc. Natl. Acad. Sci. U. S. A.* **107:** 18342–18347.

25. Crane, M.M. *et al.* 2012. Autonomous screening of *C. elegans* identifies genes implicated in synaptogenesis. *Nat. Methods* **9:** 977–980.

26. Furlong, E.E., D. Profitt & M.P. Scott. 2001. Automated sorting of live transgenic embryos. *Nat. Biotechnol.* **19:** 153–156.

27. Pardo-Martin, C. *et al.* 2010. High-throughput in vivo vertebrate screening. *Nat. Methods* **7:** 634–636.

28. Tsai, P.S. *et al.* 2003. All-optical histology using ultrashort laser pulses. *Neuron* **39:** 27–41.

29. Briggman, K.L., M. Helmstaedter & W. Denk. 2011. Wiring specificity in the direction-selectivity circuit of the retina. *Nature* **471:** 183–188.

30. Shen, E.H., C.C. Overly, A.R. Jones. 2012. The Allen Human Brain Atlas: Comprehensive gene expression mapping of the human brain. *Trends Neurosci.* **35:** 711–714.

31. Denk, W. & H. Horstmann. 2004. Serial block-face scanning electron microscopy to reconstruct three-dimensional tissue nanostructure. *PLoS Biol.* **2:** e329.

32. Lein, E.S. *et al.* 2007. Genome-wide atlas of gene expression in the adult mouse brain. *Nature* **445:** 168–176.

33. Ragan, T. *et al.* 2012. Serial two-photon tomography for automated *ex vivo* mouse brain imaging. *Nat. Methods* **9:** 255–258.

34. Yamamoto, J. & M.A. Wilson. 2008. Large-scale chronically implantable precision motorized microdrive array for freely behaving animals. *J. Neurophysiol.* **100:** 2430–2440.

35. Cham, J.G. *et al.* 2005. Semi-chronic motorized microdrive and control algorithm for autonomously isolating and maintaining optimal extracellular action potentials. *J. Neurophysiol.* **93:** 570–579.

36. Fee, M.S. 2000. Active stabilization of electrodes for intracellular recording in awake behaving animals. *Neuron* **27:** 461–468.

37. Chan, S., J. Bernstein & E. Boyden. 2010. Scalable fluidic injector arrays for viral targeting of intact 3-D brain circuits. *J. Vis. Exp.* **35:** 1489.

38. Muthuswamy, J., S. Anand & A. Sridharan. 2011. Adaptive movable neural interfaces for monitoring single neurons in the brain. *Front Neurosci.* **5:** 94.

39. Beasley, R.A. 2012. Medical robots: current systems and research directions. *J. Robot.* **2012:** 14.

40. Marescaux, J. *et al.* 2001. Transatlantic robot-assisted telesurgery. *Nature* **413:** 379–380.

41. Yuen, S.G., D.P. Perrin, N.V. Vasilyev, *et al.* 2010. Force tracking with feed-forward motion estimation for beating heart surgery. *IEEE Trans. Robot.* **26:** 888–896.

42. Sutherland, G.R., S. Wolfsberger, S. Lama, *et al.* 2013. The evolution of neuroArm. *Neurosurgery* **72:** A27–A32.

43. Hamill, O.P., A. Marty, E. Neher, *et al.* 1981. Improved patch-clamp techniques for high-resolution current recording from cells and cell-free membrane patches. *Pflugers Arch.* **391:** 85–100.

44. Eberwine, J. *et al.* 1992. Analysis of gene expression in single live neurons. *Proc. Natl. Acad. Sci. U. S. A.* **89:** 3010–3014.

45. Van Gelder, R.N. *et al.* 1990. Amplified RNA synthesized from limited quantities of heterogeneous cDNA. *Proc. Natl. Acad. Sci. U. S. A.* **87:** 1663–1667.

46. Sucher, N.J. & D.L. Deitcher. 1995. PCR and patch-clamp analysis of single neurons. *Neuron* **14:** 1095–1100.

47. Crochet, S., J.F. Poulet, Y. Kremer & C.C. Petersen. 2011. Synaptic mechanisms underlying sparse coding of active touch. *Neuron* **69:** 1160–1175.

48. Gentet, L.J., M. Avermann, F. Matyas, *et al.* 2010. Membrane potential dynamics of GABAergic neurons in the barrel cortex of behaving mice. *Neuron* **65:** 422–435.

49. Lee, A.K., J. Epsztein & M. Brecht. 2009. Head-anchored whole-cell recordings in freely moving rats. *Nat. Protocols* **4:** 385–392.

50. Lee, A.K., I.D. Manns, B. Sakmann & M. Brecht. 2006. Whole-cell recordings in freely moving rats. *Neuron* **51:** 399–407.

51. Lee, D., B.-J. Lin & A.K. Lee. 2012. Hippocampal place fields emerge upon single-cell manipulation of excitability during behavior. *Science* **337:** 849–853.

52. Harvey, C. D., F. Collman, D.A. Dombeck & D.W. Tank. 2009. Intracellular dynamics of hippocampal place cells during virtual navigation. *Nature* **461:** 941–946.

53. Kodandaramaiah, S.B., G.T. Franzesi, B.Y. Chow, *et al.* 2012. Automated whole-cell patch-clamp electrophysiology of neurons *in vivo*. *Nat. Methods* **9:** 585–587.

54. Kodandaramaiah, S.B., I., Wickersham, S.R., Bates, *et al.* 2012. Autopatcher application to single cell RNA analysis and optogenetic cell type identification. Society for Neuroscience. Online http://www.abstractsonline.com/Plan/ViewAbstract.aspx?sKey=d1660f1a-fad3-4ea9-9013-b825af0501e8&cKey=745c2c2a-313c-40a4-b762-c6bbdb091280&mKey=%7b70007181-01C9-4DE9-A0A2-EEBFA14CD9F1%7d

55. Kodandaramaiah, S.B., G. Holst, G. Talei Franzesi, *et al.* 2012. The Multipatcher: a robot for automated, simultaneous whole-cell patch-clamping of multiple neurons *in vivo*. Society for Neuroscience. Online http://www.abstractsonline.com/Plan/ViewAbstract.aspx?sKey=b4200e23-ad5c-4c94-9c7e-bfd7369b0db6&cKey=1e3bf2a4-f1fc-452e-95e4-3d639e9e3988&mKey=%7b70007181-01C9-4DE9-A0A2-EEBFA14CD9F1%7d

56. Kodandaramaiah, S.B. 2013. *Robotics for* in vivo *whole cell patch clamping*, PhD thesis. Atlanta.

57. Markram, H. 2006. The blue brain project. *Nat. Rev. Neurosci.* **7:** 153–160.

58. Alivisatos, A.P. *et al.* 2013. Nanotools for neuroscience and brain activity mapping. *ACS Nano* **7:** 1850–1866.

59. Alivisatos, A.P. *et al.* 2013. The brain activity map. *Science* **339:** 1284–1285.

60. Alivisatos, A.P. *et al.* 2012. The brain activity map project and the challenge of functional connectomics. *Neuron* **74:** 970–974.

Ann. N.Y. Acad. Sci. ISSN 0077-8923

Vision: are models of object recognition catching up with the brain?

Tomaso Poggio[1] and Shimon Ullman[2]

[1]Department of Brain and Cognitive Sciences, McGovern Institute, Massachusetts Institute of Technology, Cambridge, Massachusetts. [2]Department of Computer Science and Applied Mathematics, Weizmann Institute of Science, Rehovot, Israel

Address for correspondence: Tomaso Poggio, Massachusetts Institute of Technology, Bldg. 46-5155, 77 Massachusetts Avenue, Cambridge, MA 02139. tp@ai.mit.edu

Object recognition has been a central yet elusive goal of computational vision. For many years, computer performance seemed highly deficient and unable to emulate the basic capabilities of the human recognition system. Over the past decade or so, computer scientists and neuroscientists have developed algorithms and systems—and models of visual cortex—that have come much closer to human performance in visual identification and categorization. In this personal perspective, we discuss the ongoing struggle of visual models to catch up with the visual cortex, identify key reasons for the relatively rapid improvement of artificial systems and models, and identify open problems for computational vision in this domain.

Keywords: object recognition; visual models; supervised learning; visual cortex; feedforward; backprojection

Computer models are catching up with the brain

Object recognition is difficult

When you are watching a movie, you instantly recognize the scene and the objects in it, such as people, buildings, and cars. You may identify the location, a specific actor, and the brand of the car she is driving, her clothing, eyeglasses, wristwatch, and the like. Like other natural tasks that our brains perform effortlessly, visual recognition has turned out to be difficult to reproduce in artificial systems. In its general form, it is a highly challenging computational problem that is likely to play a significant role in eventually making intelligent machines. Not surprisingly, it is also an open and key problem for neuroscience.

Within object recognition, it is common to distinguish two main tasks: identification—for instance, recognizing a specific face among other faces—and categorization, for example, recognizing a car among other object classes. We will discuss both of these tasks below, use *recognition* to include both, and discuss later challenges beyond recognition.

Recognition in computer vision

Early computer vision recognition schemes focused primarily on the recognition of rigid three-dimensional (3-D) objects, such as machine parts, tools, and cars. This is a challenging problem because the same object can have markedly different appearances when viewed from different directions. It proved possible to deal successfully with this difficulty by using detailed 3-D models of the viewed objects, which are compared with the projected 2-D image.[1–3] These methods did not extend, however, to the recognition of more complex objects and object classes, since the variety of images of an object class such as a "dog" cannot be adequately described by the projections of a typical fixed 3-D shape. The advantage of identification over categorization in these recognition models stood in marked contrast to human vision, where general categorization is typically faster and easier than precise identification.[4]

Over the past decade or so, computational models have made significant progress in the task of recognizing natural object categories under realistic, relatively unconstrained viewing conditions. A number

doi: 10.1111/nyas.12148

Ann. N.Y. Acad. Sci. 1305 (2013) 72–82 © 2013 New York Academy of Sciences.

Figure 1. Examples of classification results for the classes "airplanes" and "bicycles." Images recognized illustrate the large variability used in current data sets used for evaluation. Classification performance for the airplane class in the 2011 competition evaluation[24] was about 88% correct recognition, using the standard recall-precision measure (88% precision at 88% recall); for the bicycles it was 75%.

of early schemes, mainly focusing on the class of human faces, obtained significant improvement over previous methods.[5–9] The techniques have evolved to reach practical applications, as evidenced by their use in current digital cameras.

The more recent versions of these computational schemes have started to deal successfully with an expanding range of complex object categories such as pedestrians, cars, motorcycles, airplanes, horses, and the like, in unconstrained natural scenes, to deal with a broad range of objects within each class.[10–20] The algorithms that were refined over the past few years can deal successfully with a large number of different object classes, in complex and highly cluttered scenes. They are being applied to databases of hundreds[21] and even thousands of object classes.[22] Yearly competitions in computer-based recognition, such as the Pascal challenge,[23,24] witness continuous improvement in the range of classes and scene complexity handled successfully by automatic object categorization algorithms.[25,26] An example of results[24] is shown in Figure 1 for the class "airplanes," illustrating the variability of images used in the test set. Comparison with human performance reveals that even the best computational schemes still fall short in most cases. In some specific tasks, however, computers have been trained to achieve high performance, including face identification, face detection, and car and pedestrian detection.[27] Moreover, recent results[28] on the very large ImageNet dataset with thousands of object classes suggest that the performance of convolutional networks,[29] when trained with very large sets of labeled examples, may begin to approach human performance, when humans are limited to short image exposure.[30] No detailed comparisons have been made, but humans are probably still superior in many recognition tasks, especially when dealing with flexible and highly variable objects and when background clutter is large. Reasons for this performance gap and possible directions for bridging the gap are discussed in later sections.

Models of the visual cortex

Over the same time period, some of the best performing recognition systems have come from research at the intersection of computational neuroscience and computer vision. Recent models of visual cortex based directly on known functional anatomy (Fig. 2),[31,32] building on a range of earlier attempts,[33–39] were able to account for and predict a number of physiological data from areas of the ventral stream from V1 and V2 to V4 and IT. This family of models was able to mimic human performance in rapid categorization tasks,[31] and some of these models of visual cortex were among the best computer vision systems at the time.[40–44]

What are the roots of computer models' recent progress?

In our view, the qualitative improvement in the performance of recognition models can be attributed to three main components. The first is the use of extensive learning in constructing recognition models. In this framework, rather than specifying a particular model, the scheme starts with a large family of possible models and uses observed examples to guide the construction of a specific model that is best suited to the observed data. The second component was the development of new forms of object representation for the purpose of categorization, based on both computational considerations and guidelines from

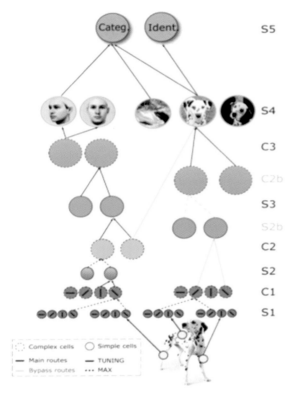

Figure 2. The hierarchical model of Serre *et al.*[31] is a quantitative version of an original proposal by Hubel and Wiesel with parameters chosen to fit the available physiological data. In the figure, S1 corresponds to a layer of simple cells in V1, and C1 to complex cells in V1. Higher layers correspond to higher cortical areas, with S4 possibly corresponding to IT and S5 to classification circuits possibly in the prefrontal cortex.

known properties of the visual cortex. These two components, representation and learning, are interrelated: initially, the class representation provides a family of plausible models, and effective learning methods are then used to construct a particular model for a novel class such as "dog" or "airplane" based on observed examples. The third component was the use of new statistical learning techniques, such as regularization classifiers (SVM and others) and Bayesian inference (such as graphical models). We next discuss each of these advances in more detail.

Learning instead of design

A conceptual advance that facilitated recent progress in object recognition was the idea of learning the solution to a specific classification problem from ex-amples, rather than focusing on the classifier design. This was a marked departure from the dominant practices at the time: instead of an expert program with a predetermined set of logical rules, the appropriate model was learned and selected from a possibly infinite set of models, based on a set of examples. The techniques used in the 1990s originated in the area of supervised learning, where image examples are provided together with the appropriate class labels (e.g., "face" or "nonface"). A comprehensive theory of the foundations of supervised learning has been developed, with roots in functional analysis and probability theory.[44–49] The formal analysis of learning continues to evolve and to contribute to our understanding of the role of learning in visual recognition.

New image representations

During learning, a recognition scheme typically extracts a set of measurements, or *features*, and uses them to construct new object representations. Objects are then classified and recognized based on their feature representation. Feature selection and object representation are crucial, because they facilitate the identification of elements that are shared by objects in the same class and support discrimination between similar objects and categories. Different types of visual features have been used in computational models in the past, ranging from simple local-image patterns such as wavelets, edges, blobs, or local-edge combinations[35] to abstract 3-D shape primitives, such as cylinders,[50] spheres, cubes, and the like.[51]

A common aspect of most earlier recognition schemes is that they used a fixed small generic set of feature types to represent all objects and classes. In contrast, recent recognition schemes use pictorial features extracted from examples, such as object fragments or patches, together with their spatial arrangement.[16,17,19,23,24,32,52] Unlike generic parts, these schemes use a large set of features, extracted from different classes of objects. The use of large feature sets is also connected to an interesting new trend in signal processing, related to overcomplete representations. Instead of representing a signal in terms of a traditional complete representation, such as Fourier components, one uses a redundant basis (such as the combination of several complete bases). This type of representation may be mirrored by the several thousands of complex shapes

Poggio & Ullman

Are models of object recognition catching up with the brain?

to which different single neurons in the posterior inferotemporal cortex of the macaque were found to be tuned.[53] Representations using such features have been used successfully in recent computer vision recognition systems for two reasons. First, these representations can be learned and used efficiently; second, they proved to effectively capture the broad range of variability in appearance within a visual class.

Two additional comments are appropriate. First, the representations described above are view-based, as opposed to object-centered models, but this should not be confused with the use of 3-D information.[54,55] A representation based on image appearance can include not only 2-D image properties, but also 3-D aspects such as local depth variations or 3-D curvature, as supported by some physiological evidence.[56–58] Second, we think that multiple object representations are likely to be used by the brain, probably for different tasks. In particular, an object-based representation using invariant 3-D properties is likely to be useful for tasks involving planning as well as perceiving object manipulation, or for slower recognition processes (e.g., those involved in mental rotation experiments, see Shepard and Metzler[59]).

New statistical learning methods

Over the past few years, the mathematics of learning has become the lingua franca of large areas of computer science and, in particular, of computer vision. As we discussed, the use of a learning framework enabled a qualitative jump in object recognition. Whereas the initial techniques used to construct useful classification models from data were quite simple, there are now more efficient algorithms originally introduced in the area of learning in the 1990s such as regularization algorithms (also called kernel machines), which include SVM[44,60,61] and boosting.[62] In addition to discriminative algorithms, the area of learning has grown to include probabilistic approaches with the goal of providing full probability distributions as solutions to object recognition tasks. These techniques are mostly Bayesian and range from graphical models[63,64] to hierarchical Bayesian models.[65–68] At the same time, the focus of research is shifting from supervised to unsupervised and semisupervised learning problems, using techniques such as manifold learning.[68] The inclusion of semisupervised problems, in which the training set consists of a large number of unlabeled examples and a small number of labeled ones, is beginning to formulate the learning problem in a more biologically plausible way, as biological organisms seem to be able to learn from experience with a surprisingly small amount of supervision. We will discuss such an approach in the context of learning invariant representations.

An interesting research direction is the formulation of learning that mirrors the combination of existing genetic information and learning from experience that is likely used by biological organisms. In this context, the range of neurobiologically plausible strategies for the computation of probabilistic graphical models capable of taking into account specific priors is quite open. As an example, it has been suggested that sampling techniques, such as MCMC, could be implemented by circuits of noisy neurons. Probabilistic models of biological vision are likely to grow in importance, at least as phenomenological and convenient descriptions of biological information processing. These models formalize in mathematical terms how established information is combined with learning from data, even if the brain does not directly use or compute probabilities.

Computer vision and the visual cortex: fundamental differences

The recent past has shown convergence of computational schemes and brain modeling. There still are, however, major differences between models and the cortex, as well as large differences in the performance between models and the brain. We will discuss below two examples of prominent features of cortical structure that have only a minor role in current computational models.

Why hierarchies?

The organization of visual cortex is hierarchical, with features of increasing complexity represented at successive layers. Models of the visual cortex have naturally adopted hierarchical structures. In contrast, in computer vision, the large majority of current schemes are nonhierarchical. Some recent models[28] have adopted a hierarchical structure and obtained high recognition performance. In addition, some computational schemes are implicitly hierarchical and possibly derive some of their power from their hierarchical organization. For instance, the so-called SIFT[19] (scale-invariant feature transform) and similar representations used in computer vision can be regarded as a three-layer network with the output roughly corresponding to intermediate units in a hierarchical cortical model.[36] What is the possible advantage of hierarchical visual representations, and can artificial systems benefit from adopting such representations?

One possible role of feature hierarchies is the need to achieve a useful trade-off between selectivity to complex patterns and sufficient tolerance for changes in position and scale, as seen in the response of IT neurons.[69–71] While scale and position invariance can be achieved quite readily in computer vision systems by sequentially scanning the image at different positions and scales, such a strategy appears unlikely to be realized in neural hardware. When properly measured, scale and position tolerance for new objects are less than originally claimed,[72] but still substantial:[70,73] for at least some of the cells in AIT, position tolerance is on the order of 2–4° in the fovea and scale invariance is on the order of a factor of 2–4, which is remarkably large. We will discuss the issue of invariant representations in the next section.[74]

A second possible advantage of hierarchical representations has to do with efficiency: computational speed and use of computational resources. For instance, hierarchy may increase the efficiency of dealing with multiple classes in parallel, by allowing the use of shared features at multiple levels. An increase in efficiency may also be related to the issue of *sample complexity*. Hierarchical architectures in which each layer is adapted through learning to properties of the visual world may reduce the complexity of the learning task and thus the overall number of labeled examples required for training. Finally, hierarchies also offer an advantage in not only obtaining recognition of the object as a whole, but also in recognizing and localizing parts and subparts at multiple levels, such as a face together with the eyes, nose, mouth, eyebrow, nostril, upper lip, and the like.[75]

Learning from very few labeled examples

One of the striking differences between current machine learning algorithms and learning in biological organisms is that the former require large amounts of labeled training data, whereas biological organisms seem to work well with a much more limited set of labeled examples. We believe that this is due in part to the ability of primate visual systems to compute representations of images that are invariant to the most common image transformations such as translations, changes of scale, rotations, changes in illumination, and, in some cases, changes in pose. One of us has proposed that the main computational goal of the ventral stream in the visual cortex is to learn during development how objects transform and then to compute *signatures* of new images that are automatically invariant to the same transformations.[76] A biologically plausible way to learn and compute invariant signatures leads to a theory of the ventral stream and of similar hierarchical architectures that achieves the dual goals of learning representations that are invariant to transformations learned in an unsupervised way and considerably reducing the sample complexity of the recognition problem, allowing categorization of new images with a small number of supervised examples. The theory suggests that invariances shape the architecture of the ventral stream and the tuning properties of its neurons. It also formalizes the main properties of neural network architectures such as the hierarchical model and X (HMAX) and convolutional networks, proposing in addition how such an architecture can learn transformations from unsupervised visual experience, instead of being hardwired for some of them.

We suggest, based on this analysis, that present high-performance computer vision systems require millions of labeled examples to achieve good performance because they do not use invariant representations. It seems that at the moment even convolutional networks take only partial advantage of invariance to transformations. It is likely, however, that invariant representations are not the only reason for reducing the number of labeled examples required for a certain level of performance. A second mechanism is sometimes called by a number of different terms, including cross-class generalization, learning to learn, and transfer learning. The basic idea in this approach is to use experience obtained during the learning of a given class to subsequently accelerate the learning of related classes. A third mechanism, more difficult to define precisely at this point but probably important, consists of *priors* (we use the term *prior* in a general way here, without necessarily implying a probabilistic interpretation)

Poggio & Ullman

Are models of object recognition catching up with the brain?

incorporated by evolution, which constrain the hypothesis space that has to be explored during recognition.

Feedforward versus backprojections

A feedforward architecture from V1 to prefrontal cortex, in the spirit of the Hubel and Wiesel simple–complex hierarchy, seems to account for several properties of visual cells. In particular, recent readout experiments measuring information that could be read from populations of IT cells[73] confirm previous estimates that after about ~100 ms from onset of the stimulus, performance of the readout classifier was essentially at its asymptotic performance during passive viewing. In addition, feedforward models also appear to account for recognition performance of human subjects for images flashed briefly and followed by a mask.[31,77] The evidence suggests therefore that a feedforward process is sufficient for a fast initial recognition phase, during which primates can already complete difficult recognition tasks involving what the image is. What, then, is the role of the extensive anatomical backprojections in the primate visual system?

Their role may be restricted to learning, but we believe that it is broader. We suggest that even when the feedforward projections by themselves may be capable of answering the *what* question of vision, the backprojections must be added to answer other questions that may be asked in a visual task, including *what is where* (as described by Marr[78]). This proposal is similar to previous ideas suggesting that visual cortex follows a hypothesis-and-verification strategy[79,80] or a Bayesian inference procedure in which top-down priors are used to compute a set of mutually consistent conditional probabilities at various stages of the visual pathway.[81] A recent model[75] along these lines demonstrated the use of initial classification using a bottom-up sweep, followed by precise localization of the object and its parts and subparts by a top-down pass.

Top-down pathways in the visual cortex also include the dorsal stream and connections between the dorsal and the ventral stream that are likely to be involved in attentional effects. A Bayesian model[82] that takes into account these bottom-up and top-down signals performs well in recognition tasks and predicts some of the characteristic psychophysical and physiological properties of attention. For natural images, the top-down signal improves object recognition performance and predicts human eye fixations, and probably attentional shifts. It is likely to be important for recognition in significant clutter, since the performance of feedforward models decreases significantly when there are more than three to five objects.[31] The top-down flow of information combined with a hierarchical representation allows the system to answer not only the what question—that is, to perform object identification and categorization—but to answer the what is where question—that is, identification and localization at multiple levels. Of course, we do not believe that the process of vision can be fully characterized in terms of answering, what is where?[78] For example, humans can recognize subtle aspects of actions, goals, and social interactions at a level which is far beyond the capabilities of our present algorithms. They can also answer essentially any reasonable question, beyond what and where, on any given image, in a kind of Turing test for vision. The top-down pathway is likely to play an important role for this broader range of visual tasks. We think therefore that while the forward pathway may be described in terms of feedforward architectures in the spirit of Hubel and Wiesel, HMAX, and convolutional networks, the full process of vision also requires verification and interpretation stages that may be best described in terms of top-down specialized routines.

Discussion: future

The more we learn about vision, the more questions appear. In this final discussion, we briefly consider two problem domains for future studies. The first focuses on how to close the gap between computer and human vision in the tasks considered above—object categorization and identification. The second part considers broader aspects of vision and its roots in evolution.

Closing the performance gap

One general question regarding possible improvements in visual recognition is whether recognition is obtained by multiple specialized mechanisms, or by a uniform scheme applied to different recognition tasks. For example, suggestions have been made that general categorization and individual recognition may be subserved by different mechanisms, or that face recognition may depend on special mechanisms not used for other object categories. It appears to us that the underlying computational

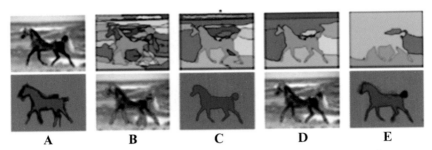

A   B   C   D   E

Figure 3. Bottom-up and top-down segmentation in computer vision. Top left: input image. Top row: bottom-up segmentation of the input image, performed at three resolutions. Bottom row: top-down segmentation. (A) Predicted figure based on stored horse model. (B) Figure outline over the image. (C) Combined bottom-up and top-down segmentation. (D) Combined figure outline over the image.

problems in different recognition tasks are similar, and can therefore be approached by the same general scheme, applied to different training sets (and possibly implemented by more than a single neuronal mechanism). The basic recognition scheme could be augmented, however, by specialized mechanisms, dealing with special cases and exceptions. The full system could then be a combination of a scheme that may be characterized as rule based, which can capture the main properties of a category and generalize broadly to novel examples, and a memory-based recognition scheme, which can deal with atypical cases and exceptions to the rule-based scheme.

We next consider future directions that we think could play a useful role in bringing the performance of artificial recognition models closer to the performance level of human vision. These are not the only possible routes for closing the performance gap, but they provide examples of promising general directions motivated by human perception that could usefully be incorporated into artificial systems.

Continuous learning of rich models

In current computational schemes, a model for the object or category of interest is constructed during a learning stage and then used for recognition. In contrast, the primate visual system exhibits continuous plasticity and can continue to learn when confronted with new examples. The disadvantage of a fixed limited training stage is that the resulting object model may remain too simple. A visual category often contains a core of typical examples, but also a large number of possible variations, atypical members, and counter-examples. An object can be recognized by its overall shape, but also by small

distinguishing parts, and both aspects need to be included in its representation. To achieve human-level performance, it appears therefore that it will be necessary to construct rich object models, learned continuously from a large number of examples.

Such use of continuous learning raises interesting computational challenges: new methods will be required to learn from errors and to continuously modify an existing representation based on new incoming information, possibly combining the rule-based and memory-based mechanisms mentioned above.

Integrating segmentation and recognition

Recognition and segmentation are related tasks in the perception of objects: we can usually recognize the object and at the same time identify in the image the precise region containing the object of interest. Historically, segmentation and recognition were treated in computer vision as sequential processes: figure-ground segmentation first identifies in the image a region likely to correspond to a single object;[83] recognition processes are subsequently applied to the selected region to identify the segmented object. More recently, computational models have started to treat the two tasks together, performing object segmentation not only in a bottom-up manner based on image properties, but also in a top-down manner based on object representations stored in memory,[84] as illustrated in Figure 3. This led to substantial progress in object segmentation; however, most current recognition systems do not include segmentation as an integral part of the recognition process. It seems to us that recognition and segmentation are closely linked tasks, and their solutions constrain each other. This

Figure 4. Challenges to future visual recognition include (A) actions performed by agents and (B) complex scene interpretation. These broader aspects of visual recognition cannot be efficiently handled by simple extension of existing recognition methods.

integration appears to be supported by considerable physiological and psychophysical evidence.[85,86] A closer integration of recognition and segmentation at both the object and part levels is likely therefore to improve the recognition of objects and their parts.

A greater challenge: vision and evolution

The brain uses vision, together with other senses, to obtain knowledge about the world and to act upon it. This knowledge goes beyond object recognition and categorization: vision is also used, for example, to recognize actions performed by agents in the surrounding environment as well as their goals and social interactions, and to complete scene understanding (Fig. 4).

It seems to us that these broader aspects of visual recognition cannot be efficiently handled by simple extension of existing recognition methods. It is likely that in addition to the general learning mechanisms currently used in object recognition models, the brain also uses specialized mechanisms, which have evolved to focus on and extract information required for making judgments about actions, goals, social interactions, and the like.

Innate structures and circuits in the brain do not by themselves incorporate full solutions to these challenging problems, but are more likely to provide useful constraints and initial biases which later lead, guided by learning from the environment, to powerful specific mechanisms.[27]

A general broad question for future studies is therefore the nature of the innate machinery used by the visual system, its genetic encoding, and how the combination of innate machinery and learning from the environment leads to our understanding of the visual world.

Acknowledgments

We thank J. Leibo and J. Mutch for reading the manuscript and for discussions. S.U. was supported by ERC Advanced Grant 269627 "Digital Baby." Additional support was provided by SkyTech, ASFOR, and the Eugene McDermott Foundation. We would like to thank the McGovern Institute for Brain Research for their support.

Conflicts of interest

The authors declare no conflicts of interest.

References

1. Lowe, D.G. 1987. Three-dimensional object recognition from single two-dimensional images. *Artif. Intell.* **31:** 355–395.
2. Grimson, W.E.L. 1990. *Object Recognition by Computer: The Role of Geometric Constraints.* Cambridge, MA: MIT Press.
3. Ullman, S. & R. Basri. 1991. Recognition by linear combinations of models. *IEEE Trans. Pattern Anal.* **13:** 992–1006.
4. Rosch, E., C.B. Mervis, W.D. Gray, et al. 1976. Basic objects in natural categories. *Cogn. Psychol.* **8:** 382–439.
5. Turk, M. & A. Pentland. 1991. Eigenfaces for recognition. *J.Cogn. Neurosci.* **3:** 71–86.
6. Brunelli, R. & T. Poggio. 1993. Face recognition: features versus templates. *IEEE Trans. Pattern Anal.* **15:** 1042–1052.
7. Sung, K.K. & T. Poggio. 1998. Example-based learning for view-based human face detection. *IEEE Trans. Pattern Anal.* **20:** 39–51.
8. Viola, P. & M. Jones. 2001. Robust real-time object detection. *Int. J. Comput. Vis.* **57:** 137–154.
9. Rowley, H.A., S. Baluja & T. Kanade. 1998. Neural network-based face detection. *IEEE* **20:** 23–38.
10. Papageorgiou, C. & T. Poggio. 1999. Trainable pedestrian detection. *IEEE* **4:** 35–39.
11. Papageorgiou, C.P., M. Oren & T. Poggio. 1998. A general framework for object detection. *Proc. CVPR IEEE* **6:** 555–562.
12. Sali, E. & S. Ullman. 1999. Combining class-specific fragments for object classification. *BMVC* **1:** 203–213.
13. Weber, M., M. Welling & P. Perona. 2000. Towards automatic discovery of object categories. *IEEE Comput. Soc.* **2:** 101–108.
14. Mohan, A., C. Papageorgiou & T. Poggio. 2001. Example-based object detection in images by components. *IEEE Trans. Pattern Anal.* **23:** 349–361.
15. Agarwal, S. & D. Roth. 2002. Learning a sparse representation for object detection. *Computer* **4:** 1–15.
16. Ullman, S., M. Vidal-Naquet & E. Sali. 2002. Visual features of intermediate complexity and their use in classification. *Nat. Neurosci.* **5:** 682–687.
17. Leibe, B. & B. Schiele. 2003. Interleaved object categorization and segmentation. *BMVC* **2:** 759–768.
18. Fei-fei, L., R. Fergus & P. Perona. 2003. A Bayesian approach to unsupervised one-shot learning of object categories. *Proc. CVPR IEEE* **2:** 1134–1141.
19. Lowe, D.G. 2004. Distinctive image features from scale-invariant keypoints. *Int. J. Comput. Vis.* **60:** 91–110.
20. Zhang, J. & A. Zisserman. 2006. "Dataset issues in object recognition." In *Toward Category-Level Object Recognition.* J. Ponce, M. Hebert, C. Schmid & A. Zisserman, Eds. New York: Springer.
21. Fei-Fei, L., R. Fergus & P. Perona 2004. Learning generative visual models from few training examples: an incremental {B}ayesian approach tested on 101 object categories. *Proceedings of Workshop on Generative-Model Based Vision.*
22. Deng, J., W. Dong, R. Socher, *et al.* 2009. ImageNet: a large-scale hierarchical image database. *Proc. CVPR IEEE* **1:** 248–255.
23. Ponce, J. 2006. *Toward Category-Level Object Recognition.* New York: Springer.
24. The PASCAL visual object classes homepage. [Online]. http://pascallin.ecs.soton.ac.uk/challenges/VOC./. Accessed 18 January 2013.
25. Felzenszwalb, P., D. McAllester & D. Ramanan. 2008. A discriminatively trained, multiscale, deformable part model. *Proc. CVPR IEEE* **8:** 1–8.
26. Vedaldi, A., V. Gulshan, M. Varma & A. Zisserman. 2009. Multiple kernels for object detection. In *Proceedings of IEEE 12th International Conference on Computer Vision 2009 (ICCV)*, vol. 1, pp. 606–613.
27. http://www.mobileye.com/. Accessed September 1, 2012.
28. LeCun, Y., B. Boser, J.S. Denker, *et al.* 1989. Backpropagation applied to handwritten zip code recognition. *Neural Comput.* **1:** 541–551.
29. Krizhevsky, A., I. Sutskever & G. Hinton. 2012. ImageNet classification with deep convolutional neural networks. *Adv. Neural Inf.* **25:** 1106–1114.
30. ArXiv: Computer Vision and Pattern Recognition, [Online]. http://www.science.io/paper/1388351/the-neural-representation-benchmark-and-its. Accessed January 18, 2013.
31. Serre, T., A. Oliva & T. Poggio. 2007. A feedforward architecture accounts for rapid categorization. *Proc. Natl. Acad. Sci. USA* **104:** 6424–6429.
32. Serre, T., M. Kouh, C. Cadieu, *et al.* 2005. A theory of object recognition: computations and circuits in the feedforward path of the ventral stream in primate visual cortex. http://dspace.mit.edu/handle/1721.1/36407. Accessed September 1, 2012.

Poggio & Ullman

Are models of object recognition catching up with the brain?

33. Fukushima, K. 1975. Cognitron: a self-organizing multilayered neural network. *Biol. Cybern.* **20:** 121–136.

34. Wallis, G. & E.T. Rolls. 1997. A model of invariant object recognition in the visual system. *Prog. Neurobiol.* **51:** 167–194.

35. Mel, B.W. 1997. SEEMORE: combining color, shape, and texture histogramming in a neurally inspired approach to visual object recognition. *Neural Comput.* **9:** 777–804.

36. Riesenhuber, M. & T. Poggio. 1999. Hierarchical models of object recognition in cortex. *Nat. Neurosci.* **2:** 1019–1025.

37. Thorpe, S. 2002. Ultra-rapid scene categorization with a wave of spikes.*Lect. Notes Comput. SC.* **2525:** 1–15.

38. Amit, Y. & M. Mascaro. 2003. An integrated network for invariant visual detection and recognition. *Vis. Res.* **43:** 2073–2088.

39. Wersing, H. & E. Körner. 2003. Learning optimized features for hierarchical models of invariant object recognition. *Neural Comput.* **15:** 1559–1588.

40. Serre, T., G. Kreiman, M. Kouh, *et al.* 2007. A quantitative theory of immediate visual recognition. *Prog. Brain Res.* **165:** 33–56.

41. Mutch, J. & D.G. Lowe. 2006. Multiclass object recognition with sparse, localized features. *Proc. CVPR IEEE* **1:** 11–18.

42. Mutch, J. & D.G. Lowe. 2008. Object class recognition and localization using sparse features with limited receptive fields. *Int. J. Comput. Vis.* **80:** 45–57.

43. Pinto, N., D. Doukhan, J.J. DiCarlo & D.D. Cox. 2009. A high-throughput screening approach to discovering good forms of biologically inspired visual representation. *PLoS Comput. Biol.* **5:** 12.

44. Vapnik, V.N. 1998. *Statistical Learning Theory.* New York: Wiley.

45. Devroye, L., L. Györfi & G. Lugosi. 1996. *A Probabilistic Theory of Pattern Recognition.* New York: Springer.

46. Cucker, F. & S. Smale. 2001. On the mathematical foundations of learning. *Bull. Amer. Math. Soc.* **39:** 1–50.

47. Poggio, T. & S. Smale. 2003. The mathematics of learning: dealing with data. *Not Am. Math. Soc.* **50:** 537–544.

48. Bousquet, O. 2004. Introduction to statistical learning theory. *Biol. Cybern.* **3176:** 169–207.

49. Poggio, T., R. Rifkin, S. Mukherjee & P. Niyogi. 2004. General conditions for predictivity in learning theory. *Nature* **428:** 419–422.

50. Marr, D. & H.K. Nishihara. 1978. Representation and recognition of the spatial organization of three-dimensional shapes. *Proc. R. Soc. B* **200:** 269–294.

51. Biederman, I. 1985. Human image understanding: recent research and a theory. *Comput. Vis. Graph.* **32:** 29–73.

52. Belongie, S., J. Malik & J. Puzicha. 2002. Shape matching and object recognition using shape contexts. *IEEE Trans. Pattern Anal.* **24:** 509–522.

53. Kobatake, E. & K. Tanaka. 1994. Neuronal selectivities to complex object features in the ventral visual pathway of the macaque cerebral cortex. *J. Neurophysiol.* **71:** 856–67.

54. Bülthoff, H.H., S.Y. Edelman & M.J. Tarr. 1994. How are three-dimensional objects represented in the brain? *Cereb. Cortex* **5:** 247–260.

55. Tarr, M.J. & H.H. Bülthoff. 1998. Image-based object recognition in man, monkey and machine. *Cognition* **67:** 1–20.

56. Janssen, P., R. Vogels & G.A. Orban. 2000. Selectivity for 3D shape that reveals distinct areas within macaque inferior temporal cortex. *Science* **288:** 2054–2056.

57. Tanaka, K. 2000. Curvature in depth for object representation. *Neuron* **27:** 195–196.

58. Yamane, Y., E.T. Carlson, K.C. Bowman, *et al.* 2008. A neural code for three-dimensional object shape in macaque inferotemporal cortex. *Nat. Neurosci.* **11:** 1352–1360.

59. Shepard, R.N. & J. Metzler. 1971. Mental rotation of three-dimensional objects. *Science* **171:** 701–703.

60. Wahba, G. 1990. *Spline Models for Observational Data.* Philadelphia: Society for Industrial and Applied Mathematics.

61. Poggio, T. & F. Girosi. 1990. Networks for approximation and learning. *Proc. IEEE* **78:** 1481–1497.

62. Freund, Y. & R.E. Schapire. 1997. A decision-theoretic generalization of on-line learning and an application to boosting. *J. Comput. Syst. Sci.* **55:** 119–139.

63. Jordan, M.I. 2004. Graphical models. *Stat. Sci.* **19:** 140–155.

64. Bienenstock, E., S. Geman & D. Potter. 1998. Compositionality, MDL Priors, and Object Recognition. *In Advances in Neural Information Processing Systems* **9:** 838–844.

65. Teh, Y.W., M.I. Jordan, M.J. Beal & D.M. Blei. 2006. Hierarchical dirichlet processes. *J. Amer. Stat. Assoc.* **101:** 1566–158.

66. Kemp, C. & J.B. Tenenbaum. 2008. The discovery of structural form. *Proc. Natl. Acad. Sci. USA* **105:** 10687–10692.

67. Sudderth, E.B. & M.I. Jordan. 2008. Shared segmentation of natural scenes using dependent pitman-yor processes. *Electr. Eng.* **21:** 1–10.

68. Belkin, M. & P. Niyogi. 2004. Semi-supervised learning on riemannian manifolds. *Mach. Learn.* **56:** 209–239.

69. Logothetis, N.K., J. Pauls, H.H. Bülthoff & T. Poggio. 1994. View-dependent object recognition by monkeys. *Curr. Biol.* **4:** 401–414.

70. Logothetis, N. K., J. Pauls & T. Poggio. 1995. Shape representation in the inferior temporal cortex of monkeys. *Curr. Biol.* **5:** 552–563.

71. Logothetis, N.K. & D.L. Sheinberg. 1996. Visual object recognition. *Ann. Rev. Neurosci.* **19:** 577–621.

72. Bruce, C., R. Desimone & C.G. Gross. 1981. Visual properties of neurons in a polysensory area in superior temporal sulcus of the macaque. *J. Neurophysiol.* **46:** 369–384.

73. Hung, C.P., G. Kreiman, T. Poggio & J.J. DiCarlo. 2005. Fast readout of object identity from macaque inferior temporal cortex. *Science* **310:** 863–866.

74. Smale, S., L. Rosasco, J. Bouvrie, *et al.* 2009. Mathematics of the neural response. *Found. Comput. Math.* **10:** 67–91.

75. Epshtein, B., I. Lifshitz & S. Ullman. 2008. Image interpretation by a single bottom-up top-down cycle. *Proc. Natl. Acad. Sci. USA* **105:** 14298–14303.

76. Poggio, T. 2011. The computational magic of the ventral stream: towards a theory. *Nature Precedings.* http://dx.doi.org/10.1038/npre.2011.6117.1 Accessed January 18, 2013.

77. Thorpe, S., D. Fize & C. Marlot. 1996. Speed of processing in the human visual system. *Nature* **381:** 520–522.

78. Marr, D. 1982. *Vision: A Computational Investigation into the Human Representation and Processing of Visual Information.* San Francisco: W.H. Freeman.

79. Carpenter, G.A. & S. Grossberg. 1987. A massively parallel architecture for a self-organizing neural pattern recognition machine. *Comput. Vis. Graph.* **37:** 54–115.

80. Hawkins, J. & S. Blakeslee. 2004. *On Intelligence.* New York: Times Books.

81. Lee, T.S. & D. Mumford. 2003. Hierarchical Bayesian inference in the visual cortex. *J. Opt. Soc. Am. A* **20:** 1434–1448.

82. Chikkerur, S., T. Serre & T. Poggio. 2009. Computer science and artificial intelligence laboratory technical report a Bayesian inference theory of attention: neuroscience and algorithms. http://dspace.mit.edu/handle/1721.1/49416. Accessed September 1, 2012.

83. Sharon, E., M. Galun, D. Sharon, *et al.* 2006. Hierarchy and adaptivity in segmenting visual scenes. *Nature* **442:** 810–813.

84. Borenstein, E. & S. Ullman. 2008. Combined bottom-up and top-down segmentation. *IEEE Trans. Pattern Anal.* **30:** 1–17.

85. Zhou, H., H.S. Friedman & R. Von Der Heydt. 2000. Coding of border ownership in monkey visual cortex. *J. Neurosci.* **20:** 6594–6611.

86. Zemel, R.S., M. Behrmann, M.C. Mozer & D. Bavelier. 2002. Experience-dependent perceptual grouping and object-based attention. *J. Exp. Psychol. Hum.* **28:** 202–217.

87. Ullman, S., D. Harari & N. Dorfman. 2012. From simple innate biases to complex visual concepts. *Proc. Natl. Acad. Sci. USA.* **109:** 18215–18220.

Ann. N.Y. Acad. Sci. ISSN 0077-8923

ANNALS OF THE NEW YORK ACADEMY OF SCIENCES

Issue: *Cracking the Neural Code: Third Annual Aspen Brain Forum*

Long-range connectomics

Saad Jbabdi[1] and Timothy E. Behrens[1,2]

[1]FMRIB Centre, University of Oxford, Oxford, United Kingdom. [2]Wellcome Trust Centre for Neuroimaging, Institute of Neurology, University College London, London, United Kingdom

Address for correspondence: Saad Jbabdi, FMRIB Centre, University of Oxford, John Radcliffe Hospital, Oxford OX3 9DU, United Kingdom. saad@fmrib.ox.ac.uk

Decoding neural algorithms is one of the major goals of neuroscience. It is generally accepted that brain computations rely on the orchestration of neural activity at local scales, as well as across the brain through long-range connections. Understanding the relationship between brain activity and connectivity is therefore a prerequisite to cracking the neural code. In the past few decades, tremendous technological advances have been achieved in connectivity measurement techniques. We now possess a battery of tools to measure brain activity and connections at all available scales. A great source of excitement are the new *in vivo* tools that allow us to measure structural and functional connections noninvasively. Here, we discuss how these new technologies may contribute to deciphering the neural code.

Keywords: brain connections; chemical tracers; tractography

Introduction

The importance of neural connections has been recognized since the beginnings of neuroscience,[1] and theories of brain function have circuitry at their heart. In particular, the past few decades have seen a resurgence of interest in studying brain connections. This is in part due to the tremendous progress that has been achieved in measuring brain connections across all scales.

Methods for measuring detailed ultrastructural and microscopic organization of neuronal networks (individual axons, dendrites, and synapses) are now entering an industrial era.[2–8] Tedious and error-prone manual delineation of intricate neural circuits over a few millimeters of tissue is currently being replaced with fast automated procedures that can process large sections of the brain. This type of high-throughput, high-fidelity data will constitute a vast wealth of connectivity information and will contribute to building a detailed understanding of neural circuits at microscopic scales.

At a larger scale, we also possess powerful tools for studying systems-level connections. In animal models, chemical tracers allow precise and accurate reconstruction of axonal bundles over their entire trajectories. In humans, modern imaging techniques allow *noninvasive* measurement of brain connections in living brains, and brought about the emerging field of *in vivo* connectomics.[9] The ability to measure brain connections in living humans has generated much excitement and triggered large concerted efforts that attempt to push the limits of these methods. One notable example is the Human Connectome Project (HCP),[10,11] a National Institutes of Health (NIH)-funded initiative that is aimed at charting the human macroconnectome in a large cohort of healthy adults using magnetic resonance imaging (MRI) and magnetoencephalography (MEG) technologies. A major focus of the HCP is to improve all aspects of data acquisition and processing to achieve much higher accuracy in building a macroconnectome than what can be achieved using current methods.[12,13]

How will these tools contribute to our understanding of brain function? Often, mechanisms of neural function are described in terms of local circuits, where the role of microconnectomics is unquestionable. For instance, microconnectomics provide statistical features and organizational principles of local connections[14–16] that can guide computational models. Macroconnectomes, on the other hand, are only beginning to play such a mechanistic role. In this paper, we review the

doi: 10.1111/nyas.12271

Ann. N.Y. Acad. Sci. 1305 (2013) 83–93 © 2013 The Authors. *Annals of the New York Academy of Sciences* published by Wiley Periodicals, Inc. on behalf of New York Academy of Sciences.

83

available tools for measuring large-scale connections, and we ask how knowledge of these long-range connections can contribute to cracking the neural code.

Measuring long-range connections

Up until the end of the 20th century, all available tools for measuring long-range connections were invasive (Fig. 1). In addition, the most accurate tools, anterograde and retrograde tracers, were (and still are) only available in nonhuman animals. Recent advances in neuroimaging are providing a new set of tools that can be used in living humans.

Tracers

Traditional methods for determining long-range connections between brain areas relied on lesion studies of axonal degradation. A revolution then occurred in the 1960s and 1970s, when a set of powerful and extremely versatile tract-tracing techniques was developed. These techniques rely on active, *in vivo* transport of compounds (e.g., proteins, amino acids, and viruses) along axons by means of cytoplasmic transport mechanisms, and are therefore extremely accurate. Tracers are injected into a source region, then after a certain amount of time, the brain is extracted, fixed, sectioned, and stained appropriately in order to detect traces of the compound at remote locations from the injection site.

A wide variety of tracers have been developed[17] (Fig. 1). These tracers differ in properties that affect their transport speed and directionality (anterograde, retrograde, or both), and whether they can cross synapses. They also differ in how they react in histochemical or immunohistochemical reactions. Certain tracers can have fluorescence properties that alleviate the need for staining. This richness and variety of available compounds means that different tracer molecules can be used simultaneously on the same animal. Several connections can be traced at once, allowing the study of detailed circuits.[18] An elegant demonstration of the power of such multiple tracer studies was shown by Lanciego *et al.*,[19] who used a combination of retrograde and anterograde tracing to ask whether pallidal afferents that reach the substantia nigra innervate neurons that project to either the caudate or the putamen. Using differentially colored staining, overlapping areas between pallidonigral afferents and different subtypes of nigrostriatal projections could easily be identified.

Depending on the tracer that is used and on the staining process, it is possible to determine not only the precise termination point of axonal projections (e.g., cortical layer), but also sometimes reconstruct the entire trajectory of axonal pathways from source to target regions.[20] In addition, modern developments in tract-tracing methods combine tracing long-range connections with detailed microanatomy.[21] By using anterograde tracing of the projections combined with immunocytochemistry to identify the postsynaptic targets, it is possible to not only determine which regions are targeted (overall) by the tracers, but also to establish fine-grained connectivity such as whether synaptic contacts are made at the target region, and to determine neuronal subtypes that are targeted by long-range connections.[21]

Tracer studies continue to provide detailed pictures of systemic connectomes in many animal models. Of particular interest are studies of nonhuman primates anatomy.[22–26] Compilations of many tracer studies in monkeys are beginning to provide quantitative data on large-scale connections throughout the cortex.[22] These large-scale connectomes are a great source of information for studying organizational principles of brain connections and guiding electrophysiological recording and interpretation in monkey studies, and also constitute an estimate or at least an approximation, of the human large-scale connectome.

Tractography

Tracers are only available in animals. As a result, and in contrast to the vast amount of connectivity data available in animal models, knowledge of human brain connectivity remains relatively poor.[27]

Studying brain connections in living humans has only been made possible following developments in diffusion magnetic resonance imaging (dMRI) in the mid to late 1990s.[28] This noninvasive technique uses the dynamics of water molecular motion as a probe of tissue microstructure. Specifically, water motion in and around biological cells is hindered by cellular processes. The directionality of this hindrance is used as an indicator of tissue orientation. For example, in a region of tightly packed axons arranged along a common average orientation, water motion is less hindered along the axons than across them. By following the motion of water, it is possible to map the orientation(s) of fibers passing through

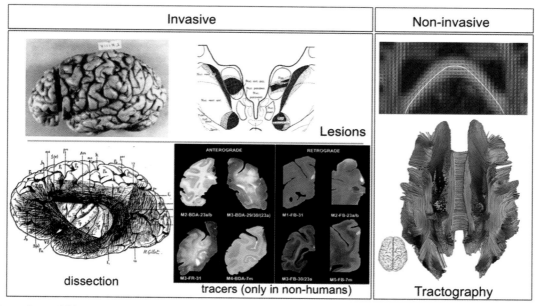

Figure 1. Available techniques for measuring anatomical connections in the brain. Lesion studies rely on Wallerian degeneration as a result of a brain lesion; the effects of the lesion can be seen postmortem at remote sites (here the thalamus) indicating the trajectories of white-matter projections (from Ref. 98). Postmortem dissections of white-matter connections date back to the 19th century (from Ref. 99). A multitude of tracers are available in animals. Shown here are example anterograde (biotinylated dextran amines, or BDA) and retrograde (Fast Blue fluorescent dye) tracers used to trace connections from the posterior cingulate cortex in macaques (from Ref. 100). The only available technique that is noninvasive is diffusion MRI tractography. The panel on the right shows how local estimates of fiber orientation, here using the diffusion tensor model, can serve to trace estimates of neural pathways. This allows reconstruction of major white-matter connections in the whole brain (top: figure from Ref. 31; bottom: image courtesy of Alexander Leemans).

each voxel of white matter. Long-range (>1 cm) connections can then be reconstructed using algorithmic approaches that integrate local estimates of fiber orientations over large distances: a technique called *tractography* or fiber tracking.[29–34]

The advent of *in vivo* dMRI tractography created a revolution in large-scale human connectomics. For the first time, we are able to virtually dissect large white-matter bundles in intact brains.[35] Tractography has two striking advantages compared with chemical tract tracing. First, it is *in vivo* (although it can also be applied *ex vivo*[36,37]), and second, it allows us to measure connections in the whole brain at once. These two features of tractography opened a large number of new research possibilities. We can now measure brain structure and function on the same brains, and thus relate structural connections to brain function and behavior,[38] analyze developmental pathways of structural connections,[39] and relate structural connections to functional segregation,[40] among many

other possibilities that were unavailable two decades ago.

Noninvasiveness and whole-brainness also come at a cost: tractography is less accurate than chemical tracing. Although sensitive to microscopic features of the tissues, dMRI produces images at a much lower resolution than microscopy (>1 mm). Information about underlying cellular processes is averaged across tens of thousands of cells or axons. Therefore, only bulk connectivity can be assessed with this technique. Furthermore, dMRI measurements of tissue orientation are indirect; actual fiber organization is only inferred from water motion, a process that can be error prone, especially when the underlying axons within an imaging voxel lack organization.[41,42] Improvements upon this promising measurement method are being carried out along several fronts. These include significant advances in imaging quality[13] and algorithm developments,[12,43–46] as well as validation and optimization using detailed comparisons of

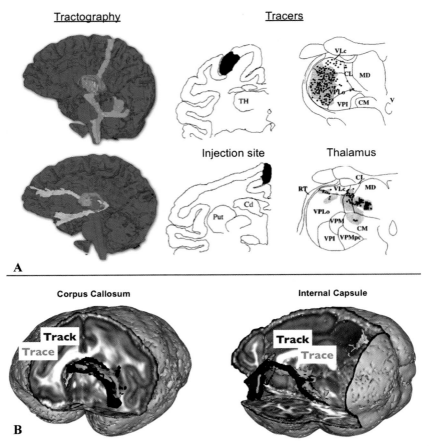

Figure 2. Two example comparisons between tractography and tracer results. (A) Connections traced from two locations in the thalamus using human dMRI tractography (left-hand side, modified from Ref. 74). Tracer studies in monkeys (right-hand side, modified from Ref. 101) shows that different thalamic regions contain traces of the injected dye depending on the cortical injection site. Comparing the two allows us to interpret the tractography result in terms of the location of the tractography seed relative to different thalamic nuclei. (B) Direct comparison of tractography and tracing of the same connections in the macaque brain. Shown here are two connections from the lateral orbitofrontal cortex traveling through the corpus callosum and the internal capsule, respectively, with a very good match between the two techniques. Modified from Ref. 47.

dMRI tractography and tracer studies in nonhuman primates.[47] Figure 2 shows two examples of comparisons between the results of tracer studies and tractography in the monkey brain.

In vivo *inference of structure from function*

Measurements of brain activity, as opposed to brain structure, can also be used as an alternative method for assessing connectivity. Resting-state functional MRI (rsfMRI; i.e., measurements of brain activity with MRI without a stimulus or an explicit task), has emerged as a powerful tool that provides information on network structure in the brain.[48] Statistical dependencies in resting-state signal (or *functional connectivity*) between remote brain areas have been shown to reflect their anatomical connections.[49–51] It is therefore thought that this type of measurement can be used, perhaps alongside dMRI tractography, to infer structural connections in the brain.[52–54]

Measurements of resting-state functional connectivity are also subject to their own biases and imperfections. Fortunately, the caveats of diffusion-based structural connectivity and resting-state functional connectivity are, to some extent, complementary. One example of this complementarity is evident when considering a caveat in tractography that is often referred to as the *distance bias*.

Connectivity strength that is usually inferred from tractography tends to decrease with distance between the source and target areas. This reflects a decrease in the certainty of the orientation measurements, which is expected from the streamlining process.[31,55] There is no such bias in rsfMRI, as the notion of functional connectivity does not rely on estimating the trajectories of the underlying axonal connections, although spatial autocorrelation in the rsfMRI signal can sometimes also induce a short-range bias in functional connectivity.

On the other hand, rsfMRI does not provide a complete picture of all anatomical connections in the brain. Clearly, anatomy constrains the statistical relations amongst neuronal time series, but this is a rather complex process and these statistical relationships are not a simple one-to-one mapping from anatomy. Statistical dependencies between connected areas may be transient (e.g., task related). Nonconnected regions may also exhibit dependencies owing to indirect projections, common input, or shared structured noise. While more sophisticated analysis methods[53] may overcome some of these limitations, it is clear that the above-mentioned rsfMRI errors are not encountered in diffusion tractography. Therefore, the two techniques have complementary weaknesses. A multi-modal approach may ultimately allow us to capitalize on their strengths, and iron out their weaknesses.

Relating long-range connections to brain function

While there is little debate that macroconnectomics is a key ingredient in understanding brain function at a systems level, it is useful to lay out specific examples of how macroconnectomes can be utilized in neuroscientific investigations. The remainder of this article highlights four broad research topics in neuroscience that will directly benefit from the availability of macroconnectomes.

Bottom-up modeling

Large-scale neuronal network simulations are increasingly used as frameworks for studying links between anatomical connections and brain dynamics. An extreme example is the ambitious BlueBrain project,[56] a colossal effort toward building a *virtual brain*, a large scale simulation on a supercomputer. Instead of summarizing small-scale activity

with simplistic models of interacting excitatory and inhibitory cells, the BlueBrain project aims to model whole macrocolumns while accounting for the great variability in cell types and their chemical properties, with temporal dynamics simulated at high resolution ($\sim$1 ms). Although such detailed bottom-up modeling promises to give us insights into local neuronal computations, we are still a long way from being able to run these types of simulations at the scale of whole brains.

At a macroscopic scale, knowledge of brain circuitry can be utilized to build computational models of large-scale networks that can generate brain activity.[57–61] Such computational models require the prespecification of a set of brain regions and precise knowledge of their connections. The online CoCoMac database[25,26] has been used as a source of such information (Fig. 3). This compilation of tracer data in macaques was used in several studies where whole brain spatiotemporal dynamics were simulated using CoCoMac data as an estimate of the underlying anatomy.[59,62–64] An example of the type of insight that these simulation studies can provide is given by Deco et al.[62] This study used CoCoMac not only to model coupling strength between brain regions, but also to provide an estimate of conduction delay. The study showed how the structural features (delay and coupling) of a simplified macaque brain network can lead to the emergence of two sets of anti-correlated oscillators consistent with many experimental observations in humans and primates.

In humans, whole-brain tractography–derived connectivity has been used as a scaffold for simulating brain activity. For instance, Honey et al.[58] used such an approach to test the extent to which we can predict resting-state fMRI measurements using systemic connectivity measures from dMRI tractography. Brain activity was generated using neural mass models of densely connected excitatory and inhibitory neurons. This local model was combined with large-scale coupling among brain areas where the coupling strength was directly proportional to structural measurements from dMRI tractography. The resulting ensemble activity was then turned into a hemodynamic signal that could be compared to empirical fMRI measurements (Fig. 3). Interestingly, this study found that brain activity over long time windows correlated strongly with the underlying anatomy.

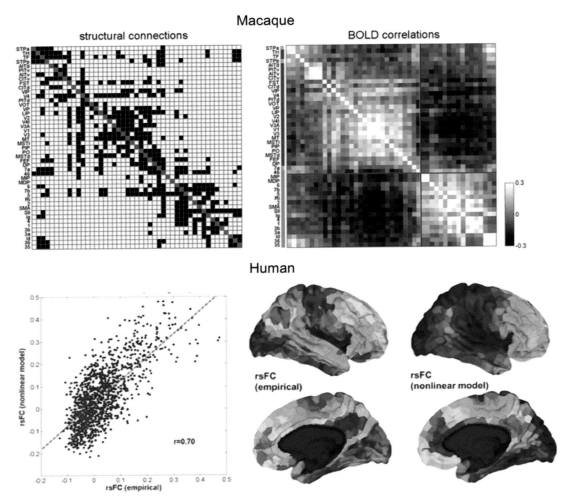

Figure 3. Bottom-up simulations of resting-state activity in macaques and humans. Top panel shows the anatomical network from CoCoMac, represented as a binary matrix of connections on the left, and the predicted fMRI functional connectivity on the right. The structure of the network induces the emergence of two anticorrelated subnetworks (modified from Ref. 64). Bottom panel shows the results of a similar modeling approach to human resting-state data. Structural connections were estimated using diffusion MRI tractography, and simulated fMRI functional connectivity was compared to empirical connectivity from a resting-state scan. The left-hand side shows the correlation between simulated and empirical whole-brain functional connectivity. The right-hand side shows a qualitative comparison between simulated and empirical functional connectivities of the posterior cingulate cortex. Modified from Ref. 58.

A notable application of such network simulation approaches is to test disconnection hypotheses by simulating lesions in large-scale network models and observing alterations of brain activity as a result of these lesions.[65,66]

Dynamic causal modeling (DCM) is another example of a set of bottom-up computational models that combine large-scale connectivity with local-scale dynamic models.[67] DCMs typically consider circuits that consist of a small set of brain regions

(3–10) and seek to model the influence that each region exerts on the other regions of the network via large scale reciprocal connections. DCM for electrophysiological data, such as electroencephalography, emphasizes detailed circuit modeling, thus enabling inferences on both large-scale interactions and local-scale properties of microcircuits.[68,69] DCMs typically require setting up an underlying anatomical model that constrains the possible routes of activity propagation between brain regions. Such a model may come from a priori anatomical knowledge, although use of dMRI tractography to constrain the anatomical model has also been suggested.[52,54]

Functional specialization

Ever since the times of Broca and his famous patient Tan, there has been overwhelming evidence for functional specialization in the brain. An important question in systems neuroscience is how to derive a subdivision of the brain that reflects this functional specialization. Neuroanatomists of the 20th century tackled this problem using postmortem histological tools that measure cytoarchitecture, myeloarchitecture, and more recently, receptoarchitecture.[70] Subsequent studies of brain function have shown that histological features are overall good predictors for functional segregation.[71] On the other hand, macroscopic landmarks, such as cortical folds, are not always good indicators for transitions between functional regions.[72] Therefore, postmortem cytoarchitectonic subdivisions cannot easily be transferred into studying living brains.

An alternative approach is to use connectivity. The extrinsic connections of a cortical area impose constraints on the type of information that an area can send or receive, and thus to some extent determine its putative function.[73] Exploiting this principle, both tractography and rsfMRI have been used to segregate gray matter according to the route of white-matter projections (extrinsic connectivity) or coherence in brain activity, respectively, both in the subcortex[74–79] and the neocortex[77,80–87] (Fig. 4). Many of these studies have shown a remarkable degree of similarity between regional borders identified using tractography and various other methods, including histological atlases, functional MRI activations, or other structural imaging modalities.

In addition to finding borders between separable regions in the brain, macroscopic connections can also help us to understand the computations and internal organization of brain regions. For instance, sensory cortices are laid out topographically, and computations within these regions are therefore spatially organized on the 2D cortical surface. Long-range connections likely reflect this topographic organization to some extent[88] and may therefore be used to further characterize the internal organization of functional regions.

Functional integration

The flip side of functional specialization is functional integration, which emphasizes how brain regions interact and influence one another. Graph theory has a central role in studying functional integration. A graph, or network, is an abstract but relatively familiar object that consists of a set of nodes and edges between these nodes. This makes it a natural mathematical description of the brain in relation to regions (nodes) and the physical connections between them (edges). Once such abstraction has been adopted, a large number of graph-theoretical concepts and measures become available for studying and quantifying the topology of the graph.[89] Many of these measures have been applied in other types of biological or social networks.

Rather than focus on the details of specific brain connections, network measures attempt to distil principles of organization and provide a set of statistics that reflect certain network characteristics. Some of these characteristic features reflect the degree to which brain regions are segregated, integrated, or clustered, highlighting, for instance, putative hubs. Other measures quantify efficiency of information propagation, establishing links between network structure and dynamics.

Numerous studies used network theory to quantify macroconnectomes derived from various types of MRI data. While there is still debate as to how various stages of data processing affect network measures, a converging picture is starting to emerge. For instance, certain brain regions of the parietal and frontal cortices have consistently been identified as central hubs connected to a structural core.[90]

Probing circuits

While a global picture of the brain macroscopic network is useful to derive general principles of organization and understand global dynamics of brain activity, scientists are often interested in studying specific brain subsystems related to specific behaviors. Such studies are of course not feasible without the ability to measure both brain function and connections in the same animal or individual.

The combination of diffusion MRI and tractography allow not only the reconstruction of major white-matter connections, but also provide measurements of microscopic and macroscopic features of those connections. For instance, certain aspects

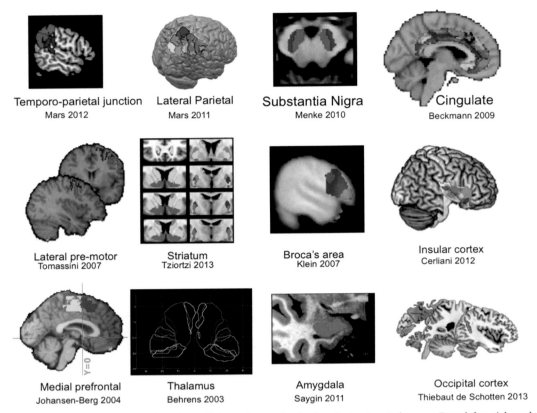

Figure 4. Examples of tractography-based parcellations of cortical and subcortical regions in humans. From left to right and top to bottom Refs. 85, 84, 102, 81, 87, 77, 103, 104, 83, 74, 105, 106, with permission.

of diffusion, such as its anisotropy, are thought to indicate axonal integrity at a microscopic level. Tract volume is another (macroscopic) measure that is often used to estimate the prominence of certain connections. Together, these micro- and macroscopic features have been used in numerous studies of brain connections in diseases,[91] development,[39] aging,[92] and a number of different behaviors such as visuospatial attention,[93] language,[94] cognitive control,[95] and skill learning.[38] Often, these studies proceed in an exploratory fashion, asking which among all measureable brain connections relate to behavior or a disease process. The availability of whole-brain connectivity afforded by tractography is therefore key to such studies.

In contrast, other studies can be guided by specific hypotheses. For instance, Aron *et al.* triangulated a cognitive-control network composed of the inferior prefrontal cortex, the subthalamic nucleus and the presupplementary motor area.[96] Their idea

was to use brain activity measurements from fMRI to determine a network of regions involved in response inhibition. They then showed that each node in the network formed connections with the other two, supporting the idea of a three-way functional–anatomical network.

Another elegant demonstration of hypothesis-driven investigation of anatomy versus function is a study by Saalmann *et al.*,[97] who were investigating the role of the pulvinar nucleus of the thalamus in selective attention. Using simultaneous recording of electrical activity in interconnected areas of the thalamus and cortex in macaque monkeys, they were able to show that the pulvinar synchronizes activity between cortical areas according to attentional allocation. In a nice demonstration of combining structural and functional measurement methods, the study used dMRI tractography to locate connected subregions of the pulvinar and cortex as a guide for electrode placement.

It is interesting to note the use of tractography in the above-mentioned study, despite the fact that the study was done in monkeys, where tracers are available and presumably more accurate than tractography. However, by using tractography, Saalman *et al.* indirectly highlight the striking advantages that tractography has over the much more accurate tracer methods available in macaques. Using tracers would have required gathering data across several animals and extrapolating the results to the animals studied with electrophysiology. Tractography provides the required connections in the same animals.

Conclusion

Our toolbox for measuring brain connections is filling up with tools that are constantly increasing in quality and accuracy. Large-scale connections can now be measured, although with some degree of uncertainty, in living brains, and we can therefore relate connections to brain dynamics and to behavior. At the same time, tremendous progress is being made in postmortem measurement of connections at microscopic scales, with a view to one day being able to map all such connections in the entire brain. The future will perhaps see these two worlds of long-range and local connections converge into a multiscale view of brain connectomics.

Conflicts of interest

The authors declare no conflict of interest.

References

1. Catani, M. *et al.* 2013. Connectomic approaches before the connectome. *Neuroimage* **80:** 2–13.
2. Briggman, K.L. & W. Denk. 2006. Towards neural circuit reconstruction with volume electron microscopy techniques. *Curr. Opin. Neurobiol.* **16:** 562–570.
3. Denk, W. & H. Horstmann. 2004. Serial block-face scanning electron microscopy to reconstruct three-dimensional tissue nanostructure. *PLoS Biol.* **2:** e329.
4. Helmstaedter, M., K.L. Briggman & W. Denk. 2011. High-accuracy neurite reconstruction for high-throughput neuroanatomy. *Nat. Neurosci.* **14:** 1081–1088.
5. Li, A. *et al.* 2010. Micro-optical sectioning tomography to obtain a high-resolution atlas of the mouse brain. *Science* **330:** 1404–1408.
6. Lichtman, J.W. & W. Denk. 2011. The big and the small: challenges of imaging the brain's circuits. *Science* **334:** 618–623.
7. Mikula, S., J. Binding & W. Denk. 2012. Staining and embedding the whole mouse brain for electron microscopy. *Nat. Methods* **9:** 1198–201.
8. Zador, A.M. *et al.* 2012. Sequencing the connectome. *PLoS Biol.* **10:** e1001411.
9. Behrens, T.E. & O. Sporns. 2012. Human connectomics. *Curr. Opin. Neurobiol.* **22:** 144–153.
10. Van Essen, D.C. & K. Ugurbil. 2012. The future of the human connectome. *Neuroimage* **62:** 1299–1310.
11. Van Essen, D.C. *et al.* 2012. The Human Connectome Project: a data acquisition perspective. *Neuroimage* **62:** 2222–2231.
12. Sotiropoulos, S.N. *et al.* 2013. Advances in diffusion MRI acquisition and processing in the Human Connectome Project. *Neuroimage* **80:** 125–143.
13. Ugurbil, K. *et al.* 2013. Pushing spatial and temporal resolution for functional and diffusion MRI in the Human Connectome Project. *Neuroimage* **80:** 80–104.
14. Hill, S.L. *et al.* 2012. Statistical connectivity provides a sufficient foundation for specific functional connectivity in neocortical neural microcircuits. *Proc. Natl. Acad. Sci. U. S. A.* **109:** E2885–E2894.
15. Druckmann, S. *et al.* 2012. A hierarchical structure of cortical interneuron electrical diversity revealed by automated statistical analysis. *Cereb Cortex* [Epub ahead of print]. DOI: 10.1093/cercor/bhs290.
16. Perin, R., T.K. Berger & H. Markram. 2011. A synaptic organizing principle for cortical neuronal groups. *Proc. Natl. Acad. Sci. U. S. A.* **108:** 5419–5424.
17. Kobbert, C. *et al.* 2000. Current concepts in neuroanatomical tracing. *Prog. Neurobiol.* **62:** 327–351.
18. Wouterlood, F.G., M. Vinkenoog & M. van den Oever. 2002. Tracing tools to resolve neural circuits. *Network* **13:** 327–342.
19. Lanciego, J.L. *et al.* 2000. Complex brain circuits studied via simultaneous and permanent detection of three transported neuroanatomical tracers in the same histological section. *J. Neurosci. Methods* **103:** 127–135.
20. Lehman, J.F. *et al.* 2011. Rules ventral prefrontal cortical axons use to reach their targets: implications for diffusion tensor imaging tractography and deep brain stimulation for psychiatric illness. *J. Neurosci.* **31:** 10392–10402.
21. Van Haeften, T. & F.G. Wouterlood. 2000. Neuroanatomical tracing at high resolution. *J. Neurosci. Methods* **103:** 107–116.
22. Markov, N.T. *et al.* 2012. A weighted and directed interareal connectivity matrix for macaque cerebral cortex. *Cereb Cortex* [Epub ahead of print]. DOI: 10.1093/cercor/bhs270.
23. Markov, N.T. *et al.* 2011. Weight consistency specifies regularities of macaque cortical networks. *Cereb Cortex* **21:** 1254–1272.
24. Yeterian, E.H. *et al.* 2012. The cortical connectivity of the prefrontal cortex in the monkey brain. *Cortex* **48:** 58–81.
25. Bakker, R., T. Wachtler & M. Diesmann. 2012. CoCoMac 2.0 and the future of tract-tracing databases. *Front Neuroinform* **6:** 30.
26. Stephan, K.E. 2013. The history of CoCoMac. *Neuroimage* **80:** 46–52.
27. Crick, F. & E. Jones. 1993. Backwardness of human neuroanatomy. *Nature* **361:** 109–110.

28. Basser, P., J. Mattiello & D.L. Bihan. 1994. Estimation of the effective self-diffusion tensor from the NMR spin echo. *J. Magn. Reson.* **103:** 247–254.

29. Basser, P. *et al.* 2000. *In vivo* fiber tractography using DT-MRI data. *MRM* **44:** 625–632.

30. Behrens, T.E. *et al.* 2007. Probabilistic diffusion tractography with multiple fibre orientations: what can we gain? *Neuroimage* **34:** 144–155.

31. Behrens, T.E. & S. Jbabdi. 2009. *MR Diffusion Tractography, in Diffusion MRI.* 333–351. San Diego: Academic Press.

32. Catani, M. *et al.* 2002. Virtual *in vivo* interactive dissection of white matter fasciculi in the human brain. *Neuroimage* **17:** 77–94.

33. Conturo, T.E. *et al.* 1999. Tracking neuronal fiber pathways in the living human brain. *Proc. Natl. Acad. Sci. U. S. A.* **96:** 10422–10427.

34. Jones, D.K. *et al.* 1999. Non-invasive assessment of axonal fiber connectivity in the human brain via diffusion tensor MRI. *Magn. Reson. Med.* **42:** 37–41.

35. Catani, M. & M. Thiebaut de Schotten 2008. A diffusion tensor imaging tractography atlas for virtual *in vivo* dissections. *Cortex* **44:** 1105–1132.

36. Miller, K.L. *et al.* 2012. Diffusion tractography of post-mortem human brains: optimization and comparison of spin echo and steady-state free precession techniques. *Neuroimage* **59:** 2284–2297.

37. Miller, K.L. *et al.* 2011. Diffusion imaging of whole, post-mortem human brains on a clinical MRI scanner. *Neuroimage* **57:** 167–181.

38. Zatorre, R.J., R.D. Fields & H. Johansen-Berg. 2012. Plasticity in gray and white: neuroimaging changes in brain structure during learning. *Nat. Neurosci.* **15:** 528–536.

39. Lebel, C., S. Caverhill-Godkewitsch & C. Beaulieu. 2010. Age-related regional variations of the corpus callosum identified by diffusion tensor tractography. *Neuroimage* **52:** 20–31.

40. Behrens, T.E. & H. Johansen-Berg. 2005. Relating connectional architecture to grey matter function using diffusion imaging. *Philos. Trans. R. Soc. Lond. B Biol. Sci.* **360:** 903–911.

41. Jbabdi, S. & H. Johansen-Berg. 2011. Tractography: where do we go from here? *Brain Connect* **1:** 169–183.

42. Jones, D. 2010. Challenges and limitations of quantifying brain connectivity *in vivo* with diffusion MRI. *Imag. Med.* **2:** 341–355.

43. Sotiropoulos, S.N., T.E. Behrens & S. Jbabdi 2012. Ball and rackets: inferring fiber fanning from diffusion-weighted MRI. *Neuroimage* **60:** 1412–1425.

44. Sotiropoulos, S.N. *et al.* 2013. RubiX: combining spatial resolutions for Bayesian inference of crossing fibres in diffusion MRI. *IEEE Trans Med Imaging* **68:** 1846–1855.

45. Jbabdi, S. *et al.* 2012. Model-based analysis of multishell diffusion MR data for tractography: how to get over fitting problems. *Magn. Reson. Med.*

46. Reisert, M. *et al.* 2011. Global fiber reconstruction becomes practical. *Neuroimage* **54:** 955–962.

47. Jbabdi, S. *et al.* 2013. Human and monkey ventral prefrontal fibers use the same organizational principles to reach their targets: tracing versus tractography. *J. Neurosci.* **33:** 3190–3201.

48. Biswal, B.B. *et al.* 2010. Toward discovery science of human brain function. *Proc. Natl. Acad. Sci. U. S. A.* **107:** 4734–4739.

49. Greicius, M.D. *et al.* 2009. Resting-state functional connectivity reflects structural connectivity in the default mode network. *Cereb Cortex* **19:** 72–78.

50. Vincent, J.L. *et al.* 2007. Intrinsic functional architecture in the anaesthetized monkey brain. *Nature* **447:** 83–86.

51. Wang, Z. *et al.* 2013. The relationship of anatomical and functional connectivity to resting-state connectivity in primate somatosensory cortex. *Neuron* **78:** 1116–1126.

52. Jbabdi, S. *et al.* 2007. A Bayesian framework for global tractography. *Neuroimage* **37:** 116–129.

53. Smith, S.M. *et al.* 2011. Network modelling methods for FMRI. *Neuroimage* **54:** 875–891.

54. Stephan, K.E. *et al.* 2009. Tractography-based priors for dynamic causal models. *Neuroimage* **47:** 1628–1638.

55. Lazar, M. & A.L. Alexander. 2003. An error analysis of white matter tractography methods: synthetic diffusion tensor field simulations. *Neuroimage* **20:** 1140–1153.

56. Markram, H. 2006. The blue brain project. *Nat. Rev. Neurosci.* **7:** 153–160.

57. Gerstner, W., H. Sprekeler & G. Deco. 2012. Theory and simulation in neuroscience. *Science* **338:** 60–65.

58. Honey, C.J. *et al.* 2009. Predicting human resting-state functional connectivity from structural connectivity. *Proc. Natl. Acad. Sci. U. S. A.* **106:** 2035–2040.

59. Honey, C.J., J.P. Thivierge & O. Sporns. 2010. Can structure predict function in the human brain? *Neuroimage* **52:** 766–776.

60. Nakagawa, T.T. *et al.* 2013. Bottom-up modeling of the connectome: linking structure and function in the resting brain and their changes in aging. *Neuroimage.*

61. Deco, G., M. Senden & V. Jirsa. 2012. How anatomy shapes dynamics: a semi-analytical study of the brain at rest by a simple spin model. *Front. Comput. Neurosci.* **6:** 68.

62. Deco, G. *et al.* 2009. Key role of coupling, delay, and noise in resting brain fluctuations. *Proc. Natl. Acad. Sci. U. S. A.* **106:** 10302–10307.

63. Deco, G., V.K. Jirsa & A.R. McIntosh. 2011. Emerging concepts for the dynamical organization of resting-state activity in the brain. *Nat. Rev. Neurosci.* **12:** 43–56.

64. Honey, C.J. *et al.* 2007. Network structure of cerebral cortex shapes functional connectivity on multiple time scales. *Proc. Natl. Acad. Sci. U. S. A.* **104:** 10240–10245.

65. Cabral, J. *et al.* 2012. Modeling the outcome of structural disconnection on resting-state functional connectivity. *Neuroimage* **62:** 1342–1353.

66. O'Reilly, J.X. *et al.* 2013. A causal effect of disconnection lesions on interhemispheric functional connectivity in rhesus monkeys. *Proc. Natl. Acad. Sci. U. S. A* **110:** 13982–13987.

67. Stephan, K.E. & A. Roebroeck. 2012. A short history of causal modeling of fMRI data. *Neuroimage* **62:** 856–863.

68. Daunizeau, J., S.J. Kiebel & K.J. Friston. 2009. Dynamic causal modelling of distributed electromagnetic responses. *Neuroimage* **47:** 590–601.

69. Kiebel, S.J. *et al.* 2009. Dynamic causal modeling for EEG and MEG. *Hum. Brain. Mapp.* **30:** 1866–1876.

70. Zilles, K. & K. Amunts 2010. Centenary of Brodmann's map—conception and fate. *Nat. Rev. Neurosci.* **11:** 139–145.

71. Amunts, K., A. Schleicher & K. Zilles. 2007. Cytoarchitecture of the cerebral cortex—more than localization. *Neuroimage* **37:** 1061–1065; discussion 1066–1068.

72. Fischl, B. *et al.* 2007. Cortical folding patterns and predicting cytoarchitecture. In *Cereb Cortex.* Charlestown, MA: Department of Radiology, Harvard Medical School.

73. Passingham, R.E., K.E. Stephan & R. Kotter. 2002. The anatomical basis of functional localization in the cortex. *Nat. Rev. Neurosci.* **3:** 606–616.

74. Behrens, T.E. *et al.* 2003. Non-invasive mapping of connections between human thalamus and cortex using diffusion imaging. *Nat. Neurosci.* **6:** 750–757.

75. Draganski, B. *et al.* 2008. Evidence for segregated and integrative connectivity patterns in the human basal ganglia. *J. Neurosci.* **28:** 7143–7152.

76. Lehericy, S. *et al.* 2004. Diffusion tensor fiber tracking shows distinct corticostriatal circuits in humans. *Ann. Neurol.* **55:** 522–529.

77. Tziortzi, A.C. *et al.* 2013. Connectivity-based functional analysis of dopamine release in the striatum using diffusion-weighted mri and positron emission tomography. *Cereb Cortex* [Epub ahead of print]. DOI: 10.1093/cercor/bhs397.

78. Devlin, J.T. *et al.* 2006. Reliable identification of the auditory thalamus using multi-modal structural analyses. *Neuroimage* **30:** 1112–1120.

79. Smith, S.M. *et al.* 2009. Correspondence of the brain's functional architecture during activation and rest. *Proc. Natl. Acad. Sci. U. S. A.* **106:** 13040–13045.

80. Anwander, A. *et al.* 2007. Connectivity-based parcellation of Broca's area. *Cereb Cortex* **17:** 816–825.

81. Beckmann, M., H. Johansen-Berg & M.F. Rushworth. 2009. Connectivity-based parcellation of human cingulate cortex and its relation to functional specialization. *J. Neurosci.* **29:** 1175–1190.

82. Eickhoff, S.B. *et al.* 2010. Anatomical and functional connectivity of cytoarchitectonic areas within the human parietal operculum. *J. Neurosci.* **30:** 6409–6421.

83. Johansen-Berg, H. *et al.* 2004. Changes in connectivity profiles define functionally distinct regions in human medial frontal cortex. *Proc. Natl. Acad. Sci. U. S. A.* **101:** 13335–13340.

84. Mars, R.B. *et al.* 2011. Diffusion-weighted imaging tractography-based parcellation of the human parietal cortex and comparison with human and macaque resting-state functional connectivity. *J. Neurosci.* **31:** 4087–4100.

85. Mars, R.B. *et al.* 2012. Connectivity-based subdivisions of the human right "temporoparietal junction area" evidence for different areas participating in different cortical networks. *Cereb Cortex* **22:** 1894–1903.

86. Rushworth, M.F., T.E. Behrens & H. Johansen-Berg. 2006. Connection patterns distinguish 3 regions of human parietal cortex. *Cereb Cortex* **16:** 1418–1430.

87. Tomassini, V. *et al.* 2007. Diffusion-weighted imaging tractography-based parcellation of the human lateral premotor cortex identifies dorsal and ventral subregions with anatomical and functional specializations. *J. Neurosci.* **27:** 10259–10269.

88. Jbabdi, S., S.N. Sotiropoulos & T.E. Behrens. 2013. The topographic connectome. *Curr. Opin. Neurobiol* **23:** 207–215.

89. Rubinov, M. & O. Sporns. 2010. Complex network measures of brain connectivity: uses and interpretations. *Neuroimage* **52:** 1059–1069.

90. Hagmann, P. *et al.* 2008. Mapping the structural core of human cerebral cortex. *PLoS Biol* **6:** e159.

91. Johansen-Berg, H. & T.E. Behrens. 2006. Just pretty pictures? What diffusion tractography can add in clinical neuroscience. *Curr. Opin. Neurol.* **19:** 379–385.

92. Salat, D.H. *et al.* 2005. Age-related changes in prefrontal white matter measured by diffusion tensor imaging. *Ann. N. Y. Acad. Sci.* **1064:** 37–49.

93. Thiebaut de Schotten, M. *et al.* 2011. A lateralized brain network for visuospatial attention. *Nat. Neurosci.* **14:** 1245–1246.

94. Catani, M. *et al.* 2007. Symmetries in human brain language pathways correlate with verbal recall. *Proc. Natl. Acad. Sci. U. S. A.* **104:** 17163–17168.

95. Neubert, F.X. *et al.* 2010. Cortical and subcortical interactions during action reprogramming and their related white matter pathways. *Proc. Natl. Acad. Sci. U. S. A.* **107:** 13240–13245.

96. Aron, A.R. *et al.* 2007. Triangulating a cognitive control network using diffusion-weighted magnetic resonance imaging (MRI) and functional MRI. *J. Neurosci.* **27:** 3743–3752.

97. Saalmann, Y.B. *et al.* 2012. The pulvinar regulates information transmission between cortical areas based on attention demands. *Science* **337:** 753–756.

98. Freeman, W. & J.W. Watts. Retrograde degeneration of the thalamus following prefrontal lobotomy. *J. Comp. Neurol.* **86:** 65–93.

99. Déjerine, J. 1895. *Anatomie des centres nerveux.* Vol. 1. Paris: Rueff et Cie.

100. Parvizi, J. *et al.* 2006. Neural connections of the posteromedial cortex in the macaque. *Proc. Natl. Acad. Sci. U. S. A.* **103:** 1563–1568.

101. Rouiller, E.M. *et al.* 1998. Dual morphology and topography of the corticothalamic terminals originating from the primary, supplementary motor, and dorsal premotor cortical areas in macaque monkeys. *J. Comp. Neurol.* **396:** 169–185.

102. Menke, R.A. *et al.* 2010. Connectivity-based segmentation of the substantia Nigra in human and its implications in Parkinson's disease. *Neuroimage* **52:** 1175–1180.

103. Klein, J.C. *et al.* 2007. Connectivity-based parcellation of human cortex using diffusion MRI: establishing reproducibility, validity and observer independence in BA 44/45 and SMA/pre-SMA. *Neuroimage* **34:** 204–211.

104. Cerliani, L. *et al.* 2012. Probabilistic tractography recovers a rostrocaudal trajectory of connectivity variability in the human insular cortex. *Hum. Brain. Mapp.* **33:** 2005–2034.

105. Saygin, Z.M. *et al.* 2011. Connectivity-based segmentation of human amygdala nuclei using probabilistic tractography. *Neuroimage* **56:** 1353–1361.

106. Thiebaut de Schotten, M. *et al.* 2012. Subdivision of the occipital lobes: an anatomical and functional MRI connectivity study. *Cortex* [Epub ahead of print]. DOI: pii: S0010-9452(12)00342-5. 10.1016/j.cortex.2012.12.007.